SACRED STREAMING?

*Pastoral Warnings Against
Digitizing Church*

SACRED STREAMING?

*Pastoral Warnings Against
Digitizing Church*

NICK THOMPSON

To Ralph English (1936–2025)

A father in the faith whose persevering presence in public worship in spite of Parkinson's taught me more about the desirability and irreplaceability of Christ's gathered people than any book or sermon ever could.

Blessed are those who dwell in your house,
ever singing your praise!
Blessed are those whose strength is in you,
in whose heart are the highways to Zion.
As they go through the Valley of Baca
they make it a place of springs;
the early rain also covers it with pools.

—Psalm 84:4–6

Published by Ezra Press

5519 Highway 153, Unit 11, Chattanooga, TN 37343

423.287.5656 | ezrapress.ca

Nick Thompson, *Sacred Streaming?: Pastoral Warnings Against Digitizing Church*

This title is also available as an Ezra Press ebook.

Visit ezrapress.ca/collections/ebooks.

Cover design by Jordan Cecile

Interior design by Steven R. Martins

Printed in the United States of America.

Unless otherwise indicated, all Scripture quotations are from the ESV® (The Holy Bible, English Standard Version®), copyright © 2001 by Crossway, a publishing ministry of Good News Publishers. Used by permission. All rights reserved.

Any internet addresses (websites, blogs, etc.) referenced in this book are provided as resources. Their inclusion does not constitute or imply an endorsement by Ezra Press, nor does Ezra Press guarantee the accuracy or content of these sites throughout the lifespan of this book.

Library of Congress Cataloguing-in-Publication Data is Forthcoming.

ISBN: 979-8-9916932-1-9

TABLE OF CONTENTS

How we worship matters. With the proliferation and increasing ubiquity of digital technology in recent decades—as well as the global spread of the COVID-19 pandemic in 2020—churches need more than ever to grapple with the nature of Christian "assembling" (Heb. 10:25) and to ask whether virtual church constitutes corporate worship. With biblical balance, wisdom, and precision, Nick Thompson helps us think biblically about technology in worship. In doing so, he not only examines the concept of 'online church,' but he also considers such fruitful topics as the second commandment, the Sabbath, the regulative principle of worship, the relationship between church and state, the nature of humanity, and the importance of embodiment. This book is thoughtful, substantive, and rich.

–JOEL R. BEEKE, Chancellor and Professor of Systematic Theology and Homiletics, Puritan Reformed Theological Seminary, Grand Rapids, Michigan

Sacred Streaming? by Nick Thompson is a timely and insightful exploration of how digital technology, particularly livestreaming, is influencing and even reshaping the church's worship and fellowship. Grounded in biblical principles and a Reformed theological perspective, Thompson challenges readers to evaluate whether some of these tools serve or undermine the purpose and special character of the sacred gathering of believers for worship. With clarity and pastoral wisdom, this book equips Christians to navigate the digital age while staying faithful to the truth of God's Word. A must-read for church leaders and believers seeking to honour true worship in an increasingly digital world.

–JOE BOOT, President & Founder of the Ezra Institute, UK

This is a provocative book that may be uncomfortable to read. You may even disagree with some of its conclusions. But you should read it anyway. It addresses the high calling of the church — to worship our Triune God and enjoy face-to-face fellowship with one another. Can these things be done virtually? What is lost when livestream worship is offered as a substitute? Considering these questions will help sharpen your understanding of the truths of scripture, and will challenge you as you consider the place of livestream technology in the life of Christ's church."

—JONATHAN MASTER, President, Greenville Presbyterian Theological Seminary, Greenville, South Carolina

In a time when everything and everyone seems to pursue a virtual "reality," Thompson in *Sacred Streaming?* provides sober-minded pastoral reflection on the dangers and limits of its use for worship and the church, reminding us of the importance of 'embodied' worship of the Triune God.

—WILL WOOD, Associated Professor of Old Testament, Reformed Theological Seminary, Atlanta

How we worship is a subject that deserves more attention than we often give it. The same is true of how we use technology, and even more so when these two subjects converge. Pastor Nick Thompson has done the church an immense service by offering insightful, careful, and biblical wisdom regarding the proper place of technology in corporate worship. I can hardly see how an American Christian living in the 21st-Century can afford not to read these pages and wrestle with their conclusions. It's that important.

—JONATHAN LANDRY CRUSE, Pastor, Community Presbyterian Church, Kalamazoo, Michigan

THE DEATH OF THE ECCLESIASTICAL EXPERIENCE?

A<small>T THE TIME OF WRITING</small>, the tech giant Apple is in the hotseat after releasing a slick advertisement for the new iPad Pro.[1] The commercial features an industrial trash compactor slowly crushing a pile of carefully-stacked creative objects. Paint cans, books, musical instruments, sculptures, camera lenses, and more are flattened as the compactor closes in upon them. When its destructive metal jaws finally open, we find that all these creative mediums have been compressed and combined into a single device—the new iPad Pro.

The message is obvious. "Who needs pianos, clay, and paper-and-ink tomes when you can have them all combined in one digital tool that fits in your backpack?" The latest iPad, we are told, has rendered old-fashioned creative mediums unnecessary, and even undesirable.

The message is not new. Apple, together with other leading tech companies, has been selling us this bill of goods from the beginning. So why was this particular advertisement met with

---

1. John Daniel Davidson, "Apple's iPad Ad Boasts Of Replacing The Real With The Fake," *The Federalist,* May 9, 2024.

such angry backlash that Apple's CEO felt compelled to issue an apology? The reason is twofold.

First, the astounding capabilities of the new iPad bring Apple closer than ever to reaching their intended goal of replacing every other creative tool. I remember when the first iPod was released, promising to replace my stacks of CDs by enabling me to house my entire music library on a pocket-sized digital device. But the original iPod didn't have the capability to compose music, craft art, or capture photography. With the new iPad, however, you can compose music without a physical guitar, craft art without physical paint, and capture photography without physical film. Digital technology has advanced to such a point that the threat of losing these treasured means of creative output is greater than ever, and people are waking up to the threat and responding accordingly.

Intimately related to that is a second reason, namely, that Apple has never so graphically communicated their intentions. It is one thing to tell people your device has software that can compose and replicate music, but it is another thing entirely to tell people your device is intended to replace the wood, strings, and brass of physical instruments. That was Apple's message—a glass touchscreen has arrived to forever replace the piano, trumpet, and guitar. Any musician not offended by such a suggestion needs to question whether he is really a musician!

Perhaps it is because I'm not a musician and don't have an artistic bone in my body, but my initial response to the commercial was gratitude. I found it to be refreshingly honest, not because what it claimed was true to reality, but because what it claimed was true to Apple's intentions. Sometimes we need a commercial like this to shed light on where we are as a society. We are living in an age in which digital technology is attempting to replace everything meaningful. That is true, not just of the creative arts, but even more importantly of the church and her worship.

Imagine that instead of including paint and pianos, this Apple advertisement featured a pulpit, a stack of Bibles and hymnals, and a communion table with bread and wine, along with flesh-and-blood Christians embracing one another, serving one another, and welcoming one another into their homes. The metal mouth of the compactor slowly chomps down on these essential ecclesiastical realities until they are compressed into an iPad Pro. The picture is nothing short of tragic.

The actor Hugh Grant, with more than a hint of sarcasm, provided commentary on Apple's commercial: "The destruction of the human experience. Courtesy of Silicon Valley."[2] Similar commentary would befit our imaginary commercial. It equals the destruction of the ecclesiastical experience. Sadly, that fictional advertisement is a graphic illustration of the role technology has come to play in the lives of many Christians today. The gathered church is now transmitted through a glowing screen.

There are various reasons for this. Some churches have exclusively embedded their life online, not even offering in-person worship and fellowship. The early twenty-first century gave rise to a new ecclesiastical breed of online and virtual "churches" located solely on a web domain. But for most congregations, livestream has not entirely replaced the gathering of the saints in flesh-and-blood proximity; instead it has become an optional alternative. Congregations now stream their gatherings for those not present. The reasons for not being present are manifold. Some are unwilling to gather, while others are

2.   https://x.com/HackedOffHugh/status/1788183871504204257.

genuinely unable to gather. Some celebrate the church being crushed into an iPad, while others resign to it as the next best thing given their physical immobility. For at least it can simulate the ecclesiastical experience, even if artificially so.

My heart grieves for those who are providentially hindered from gathering. I write as a pastor who is intimately aware of the agonizing complexities at play here, and that not everyone chooses "virtual church" out of spiritual indifference or theological ignorance. Some of the holiest saints are simply unable to leave their homes. This has been the case for thousands of years, but it was not until relatively recently that livestreaming a church service became an option. Just because something is an option, however, does not mean we should choose it, especially when it comes to something so sacred as the public worship of God. We must think very carefully about these things.

Technology's Shaping Power

Those of us living in the West have been immersed in the idea that autonomy, not community, is the path to finding ourselves. The catechism of expressive individualism begins like this:

> Question: What is the chief end of man?
>
> Answer: Man's chief end is to generate himself and to assert himself forever.

It is the cultural air we breathe. But what we often fail to grasp is that this view of self-hood has a powerful partnership with digital technology. "What's crucial to realize," writes Samuel James, "is that alongside the philosophical revolution of expressive individualism, the digital technology revolution has exploded, and in the process it has provided the revolution of expressive individualism with its most important, most enchanting, and most effective vehicle."[3]

Many Christians are quick to defend digital innovations as ethically neutral tools. They tell us that social media and livestream are neither good nor bad, and that what matters is how we use them. In his typical no-nonsense style, Marshall McLuhan labeled this view of technology "the numb stance of the technological idiot."[4] Those are strong words, but McLuhan understood that embedded in every tool we create is a vision of reality. Technology is humanity's attempt to make the world a better place, and it presupposes a certain view of the world, including what is desirable and good. Furthermore, the tools we make, especially the tools we use to communicate, shape our view of the world in a profound way. No technology is neutral.[5] Every human innovation is shaped by a view of reality, and every innovation shapes our view of reality. Many of our digital technologies have an inherent bent toward expressive individualism, and the uncritical use of them results in us being shaped after the same.

Take Facebook as an example. It is a radically individualistic platform presenting itself as communal. It provides users

---

3. Samuel James, *Digital Liturgies: Rediscovering Christian Wisdom in an Online Age* (Wheaton, IL: Crossway, 2023), 7.

4. Marshall McLuhan, *Understanding Media: The Extensions of Man* (New York: McGraw-Hill, 1964), 19.

5. John Dyer positively quotes one of his seminary professors as stating, "One of the most dangerous things you can believe in this world is that technology is neutral" [*From Garden to City: The Redeeming and Corrupting Power of Technology* (Grand Rapids, MI: Kregel, 2011), 15]. Similarly, Tony Reinke asserts, "No technology is ambivalent; each one comes with certain biases and tendencies" [*God, Technology, and the Christian Life* (Wheaton, IL: Crossway, 2022), 70].

with a digital space to craft their own self-image and to gather "friends" who will affirm them in that image. With mind-boggling algorithms, it curates each user's individual feed to give them the information they are statistically projected to want. It is no neutral tool; it is designed according to a certain view of individuality, community, and reality which makes it an extremely effective vehicle for the propagation of expressive individualism among its billions of users. That doesn't mean we can't use it, but it does mean that we need an extraordinary amount of moral fortitude and spiritual maturity to use it wisely. For our digital technologies, being shaped by a particular view of reality, shape us after that particular view of reality, often imperceptivity.

Carl Trueman, in *The Rise and Triumph of the Modern Self,* notes the role technological advance played in paving the way for the sexual revolution. What do irrigation systems, lightbulbs, antibiotics, and automobiles have to do with transgenderism? More than you might think. Irrigation meant we were no longer dependent upon the weather for a good harvest. Lightbulbs meant we no longer looked to the sun to structure our days. Antibiotics meant we no longer feared the bacteria that killed our ancestors. Automobiles meant we were no longer tethered to one geographical location. According to Trueman, "all these developments have served to weaken the authority of the natural world and persuade human beings of their power."[6] Once you are convinced that you can override nature to serve your own ends, it is not a far leap to claim you can override your biological sex to serve your psychological desires. Despite

---

6. Carl Trueman, *The Rise and Triumph of the Modern Self: Cultural Amnesia, Expressive Individualism, and the Road to Sexual Revolution* (Wheaton, IL: Crossway, 2020), 41.

popular belief, "technology affects in profound ways how we think about the world and imagine our place in it."[7]

Prior to the invention of the printing press, Christians had no access to the Scriptures apart from the local church. Prior to the invention of radio, cassette tapes, and television, Christians had no access to preaching apart from assembling with God's people on God's day. Prior to the invention of the telephone, Christians could not fellowship apart from meeting together. I'm not suggesting we throw out our personal copies of God's word, nor that we quit speaking to one another over the phone or listening to recorded sermons. But we need to recognize that with the rise of each of these technological innovations came a subtle message—"You can have the spiritual benefits of the church without gathering as the church."

What makes our moment unique is that the distinct ecclesiastical benefits mediated through each of these individual technologies have now been remarkably brought together in a single, accessible, and somewhat affordable device, giving the sense that the church has indeed been successfully compacted into a technological tool. We need to ponder how the iPad Pro and its equivalents might be shaping our religious convictions. For to alter Trueman's words a bit, *technology affects in profound ways how we think about the church and imagine our place in it.* With every technological advance, that only becomes more true.

Livestream's Shaping Power

In November of 2008 something monumental happened. YouTube hosted its first live video. The technology to stream something in real time via the internet has been around since 1995,

---

7.   Trueman, *The Rise and Triumph of the Modern Self,* 41.

but very few had the ability to use it in its beginning stages. With the explosion of internet media platforms like YouTube, however, livestream technology quickly became accessible to anyone and everyone. People found themselves able to experience what was happening at that very moment in other parts of the world, and simultaneously share with the world what they were experiencing. All that was needed was a device with a screen, a video camera, and an internet connection. It was exciting, and the masses embraced it without a second thought.

But embedded in this digital technology was a vision of reality with profound shaping power. To understand that, it is helpful to ask, "What alleged problem did livestream seek to solve?" After all, every invention sets out to fix or improve something, but determining *what* needs fixed or improved and *why* it needs fixing or improving entails a particular vision of reality. I don't pretend to know what was in the minds of the inventors of livestream, but it seems obvious that this technological medium seeks to improve human life by fixing at least three problems.

First, there is the problem of *distance*. We are finite creatures who are bound to a single place at a single time. Our spatial finitude is a blessing because it is a constant reminder that we are not God. But our unipresence is also a great hindrance, often preventing us from being where we want or need to be. Our sinful tendency is to view our spatial limitations as a curse to be overcome rather than a gift to be humbly embraced. Livestream, by way of its form, can serve that sinful tendency by bringing what is distant near and giving us the subtle perception that we have the God-like ability to be in many different places at once. Embedded in this tool is the assumption that, at least in some situations, spatial limitations are bad.

Then there is the related problem of *disconnection*. Think of the father deployed overseas. The distance results in a painful disconnect with his wife and children. He misses all the little moments to relate with them throughout a given day. But wouldn't it be great if he was able to watch his daughter's ballet recital and look into her face immediately afterward to tell her how proud he is of her even though they are physically separated by thousands of miles? Or to broaden the scope, wouldn't it be great to be able to see and hear what is going on around the world so as to not be ignorant of current events? A century ago such ideas were relegated to science-fiction novels. Livestream, along with other digital technologies, gives us the sci-fi-like ability for boundless connection with those we love and with the world. But as the most connected people ever to exist, we are also some of the most lonely people ever to exist.[8] Is that perhaps because the connection God designed us to share with one another and the world cannot be adequately experienced without physical presence and physical touch? In order to overcome our spatial limitations, we have to disregard our inherently-spatial bodies. For to be omnipresent you also need to be incorporeal (i.e., not composed of matter). Streaming technologies, with their goal of connecting humans who are physically distant, cannot help but downplay the vital importance of that which is physical and bodily.

---

8. The U.S. Surgeon General sounded the alarm in 2023 concerning what he called an "epidemic of loneliness and isolation." See Vivek H. Murthy, "Letter from the Surgeon General," in *Our Epidemic of Loneliness and Isolation: The U.S. Surgeon General's Advisory on the Healing Effects of Social Connection and Community* (Washington, DC: Office of the U.S. Surgeon General, 2023).

Finally, there is the problem of *delay*. Until the invention of the telegraph and the telephone, connecting in any way with those physically distant from us always entailed delay. You sent a hand-written letter across the ocean and waited months before receiving a response. War broke out in another country and sometimes news would not reach you until it was all over. The term *live* entails the elimination of all delay. Instantaneous connection regardless of distance. Delay can certainly be problematic in certain contexts, but it is also a good gift from God that teaches us patience and reminds us that we are time-bound creatures, not the timeless Creator. In our depravity, we come out of the womb wanting what we want and wanting it *now*. Livestream, by its very nature and name, promises to give us just that.

We are spatially-limited creatures, but livestream digitally bridges the distance. We are relationally-limited creatures, but livestream digitally bridges the disconnect. We are temporally-limited creatures, but livestream digitally bridges the delay. It does all of this by way of its form, regardless of the content being streamed.

In our fallen condition, the message we can hear this digital tool whispering to our souls is, "You can be like God" (see Gen. 3:5). That doesn't mean we cannot or should not use livestream, but it does mean we need to be very watchful of our hearts when we do. If we are not vigilant, we will be vulnerable to being slowly and subtly shaped after the serpent's lie. Livestream is not neutral. By enabling us to digitally transcend our creaturely limitations, it communicates a message that has the ability to subtly shape our sense of God, self, and the church.[9]

Digital Church in the Pandemic's Wake

As livestream technology became more accessible, churches began experimenting with it, but it was not until 2020 that this particular technology was embraced in a nearly universal manner by the church in the West. When the Covid-19 pandemic hit, church leaders found themselves under government orders to close the doors of their church buildings. The inability to meet resulted in painful distance, disconnect, and delay which they sought to alleviate by utilizing livestream. Even the rural church with thirty members began streaming worship services, prayer meetings, and interactive fellowship times. It became the new normal almost overnight.

Suddenly believers were seemingly experiencing their church's worship and even engaging in conversation and prayer with fellow church members without ever getting up off the couch. What had always been enjoyed publicly and physically was now experienced from the privacy of one's home via a screen. If the industrial crusher was in need of a final push to flatten the life of the church into a polished device, the pandemic was it.

While the far-reaching Covid lockdowns might be a thing of the past, the far-reaching use of livestream in the church is not. In 2021 Colin Hansen predicted, "Livestream technology

---

9. For a more extended exposition of the goodness of our creaturely limitations, see Nick Thompson, *Growing Downward: The Path to Christ-Exalting Humility* (Grand Rapids: Reformation Heritage Books, 2022), 13–51.

will almost certainly prevail in a majority of churches."[10] It has indeed, becoming the new norm. Go to many church websites and livestream is presented as one of two equally legitimate ways to engage in the worship of the church. Read many church signs and you will be warmly invited, "Join us online!" Considering just how massive of a shift this is (for twenty centuries the church never even considered the possibility of gathering in a non-physical way) and considering just how weighty the public worship of God is (a theme we will spend much time thinking about in the pages ahead), it is surprising how little wrestling the church has done with the theological, ecclesiastical, and practical implications of using livestream technology in its services. Could it be a sign that digital technology has altered our perception of the church more than we care to admit?

Even if we are awake to the subtle shaping power of livestream and are proactive to rejoice in our creaturely limitations as we use it, we still need to question whether it is a fitting tool for the church's worship and fellowship. What is appropriate for a football game or a graduation ceremony may be woefully unbefitting of the sacred assembly. Can church be digital? Can public worship be experienced without publicly gathering? Does livestream serve or subvert the communion of believers? I don't pretend that these are easy questions to answer, but we need to ask them. For if the Covid-lockdowns taught us anything, it was that this technological medium profoundly alters how Christians relate to God and one another. Livestream radically changes how we do church, and that should give reason for serious pause and careful consideration.

---

10. Colin Hansen, "What We Lose When We Livestream Church," *The New York Times*, August 8, 2021.

The Scary Work of Sounding the Bullhorn

Sacred Streaming? is not anti-digital, but it is structured around four warnings. That is not because I am a pessimist; it is because I am a pastor. Warning people of spiritual danger is an integral part of my prophetic calling (Gal. 5:21; Col. 1:28). As a watchman on the walls of the church, I have a responsibility to sound the alarm when I perceive deadly threats at hand (Ezek. 3:16–27; 33:1–9). Admittedly, it is a scary task to be a watchman because you don't always perceive things rightly. Weak vision, sleep deprivation, or the darkness of the night can cause you to mistake a friend for a foe. I acknowledge the very real possibility of that here. I am a mere man with fallible understanding, limited foresight, and sinful biases, and I wouldn't be surprised if some of what I say in the pages ahead requires future revision (though I sincerely hope that is not the case).

So why am I sounding an alarm? Because when a watchmen is convinced there is serious danger at hand and remains silent out of fear that he might possibly be wrong, he neglects his most basic duty. I perceive a major threat, and I feel duty-bound to raise my proverbial bullhorn. In the pages ahead I will not tell you that livestream is the church's archnemesis. But I will warn you of the dangers I see of using livestream in the church. At the end of the day, you are responsible before God to decide whether those warnings are legitimate or not and how best to respond to them.

Whenever the subject of digital technology arises, we are in deep waters. As I think of the members of my church, no two relate to digital technology in exactly the same way. Some are on social media; others are not. Some have smart phones; others do not. Some allow their children to play Xbox; others do not. Technology use is largely in the realm of wisdom, requiring the application of biblical principles to complex circumstances. We won't all come out in the same place. But what is imperative is for us to understand the inherent dangers at play, so that if we choose to use a certain technological tool, we use it wisely unto God's glory and humanity's good.

My greatest fear in writing *Sacred Streaming?* is not that I might be proven wrong. What scares me most about blowing this bullhorn is the potential it has to divide my fellow watchmen and the royal city we are called to protect. I believe livestream use desperately needs to be talked about and even debated in the church, but we cannot be at each other's throats over it. You might disagree with my argumentation at various points in the pages ahead. Please don't write me off. Help me to understand your viewpoint with an open Bible. Let's wrestle with this together, allowing our disagreements to challenge and sharpen one another, not cut and sever ties with one another. For though we might disagree on whether livestream is a legitimate and fitting medium of public worship, can we not agree that the corporate worship of God the Father through God the Son by God the Spirit is the supreme end of all created reality? Though we might disagree about which circumstances (if any) warrant the use of livestream in the church, can we not agree that no form of technology, no matter how sophisticated, will ever be able to replace the gathered church? Yes, we might disagree on exactly how to carry out God's worship in the church, but we mustn't do so at the expense of the fundamental convictions that unite us!

A Call to Digital Minimalism in the Church

Douglas Adams rather humorously articulates the error we often fall into when it comes to technology. He states that we tend to view the technology that was invented before we were born as "normal," the technology that was invented between our birth and age thirty as "incredibly exciting," and the technology invented after our thirtieth birthday as "against the natural order of things and the beginning of the end of civilization as we know it until it's been around for about ten years when it gradually turns out to be alright really."[11] It is funny because it is so true! I've reflected hard on this insight of Adams because when the lockdowns began in 2020 I was 29 years old. Was I suffering from and continuing to suffer from the technological pessimism that characterizes so many adults when confronted with innovation? Should you put this book down because it is nothing more than the boorish ramblings of a 30-something-year-old wishing we could go back to the golden days before the digital revolution ruined everything for the church? I don't think so.

To be transparent, if there was a spectrum on digital technology use spanning from Wendell Berry on the far right to Elon Musk on the far left, I would be more to the right. I am regularly made fun of for my rather strange relationship to digital technology. I have no social media accounts. My iPhone has been dumbed down with a greyscale screen, no email or internet access, no notifications, and only a few select apps. I'm notoriously slow at responding to emails and text messages. I despise e-readers and cannot fathom ever trading my overflowing bookshelves for a tablet (my deepest sympathies to those reading this on a Kindle!). I could go on, but you get the point. I bring certain biases to this discussion, and so do you. But the

---

11. Douglas Adams, "How to Stop Worrying and Learn to Love the Internet," *The Sunday Times,* August 29, 1999.

reason I relate to technology in my personal life the way that I do is because of what I value.

In his book *Digital Minimalism,* Cal Newport proposes a "philosophy of technology use in which you focus your online time on a small number of carefully selected and optimized activities that strongly support things you value, and then happily miss out on everything else."[12] Instead of allowing digital technology to overtake our lives and shape our values, we ought to limit usage to that which optimally serves what we value and willingly forgo it at every point it doesn't. The beauty of this philosophy is that it does not come with the one-size-fits-all message of, "You must use technology in this way." For example, you might have a Facebook account because you value family and find it to be an effective way to connect with distant relatives, whereas I might not use Facebook because I value family and find it to be an ineffective form of communication with distant relatives and a distraction from the wife and kids who are sitting right in front of me. We both value family, but as we apply that biblical value to our lives it leads to two very different approaches to social media.

I'm not here to legislate the precise way you and your church use or don't use digital technology. But as Christians, biblical convictions and values must be in the driver's seat, steering our technology usage. For when technology usage drives our values, it won't be long before those values cease to resemble anything that is recognizably biblical. Newport may not be writing as a follower of Christ, but I believe his minimalistic approach to technology is precisely what the church needs in this hour. As John Dyer explains, "Technology should not dictate our val-

---

12. Cal Newport, *Digital Minimalism: Choosing a Focused Life in a Noisy World* (New York: Portfolio, 2019), 28.

ues or our methods. Rather, we must use technology *out of* our convictions and values."[13] When we do that, it will limit what technologies we use, how we use them, and in what contexts we use them.

The Book In Your Hands

Because of our convictions and values, the elders of Cornerstone Presbyterian Church where I serve as pastor made the difficult decision to stop making our worship services available to the public via livestream back in May of 2023.[14] The book in your hands is the fruit of my personal wrestling with this matter before God and His word. I am an unashamedly Reformed and Presbyterian pastor, and every argument pulsates with the interpretive insights and theological heritage of Reformation and Post-Reformation thought. I purposefully quote many great Christian thinkers (as well as non-Christian ones) to make clear that there are no doctrinal novelties here. If anything is new, it is the rigorous application of those revealed truths to the narrow subject of digitizing church.

As a pastor, I realize this is a very sensitive subject that impacts real people. I will wrestle with that in the pages ahead, particularly the difficulty of ministering to the homebound.[15] But perhaps the most fundamental conviction that undergirds

---

13. Dyer, *From the Garden to the City*, 25.

14. Like so many churches, Cornerstone had begun using this technology to stream public worship in response to the Covid-19 lockdowns in early 2020. When they called me as their pastor in the summer of 2020, the congregation had returned to physically gathering, but we continued to make livestream available for the next three years.

15. By the homebound, I mean those who are physically unable to gather with the church. I distinguish that from the homebody who is physically able to gather with the church but willfully chooses not to.

all of my thinking on this matter is that *theology must drive pastoral practice*. Always. My fear is that often in this discussion we allow our love for the elderly, the sick, and the lost to drive us to conclusions that actually contradict our theological convictions and values, and in so doing we fail to really love these people as Christ calls us to.

Sacred Streaming? is not intended to legislate a detailed program for using or not using digital tools in the church. Such matters defy rigid dogmatism. Nowhere will you find me commanding you to pull the plug on livestream. Nor will I simplistically claim that there are only two possible approaches to this digital medium. A church may decide to pull the plug on livestream in certain contexts, while utilizing it in others (we'll think about this much more in the conclusion). That being said, the pages ahead attempt to set forth the relevant biblical data in order to commend to you the fittingness of pulling the plug on ecclesiastical streaming, at least in most circumstances.[16] For just because a certain technology is available and provides certain benefits doesn't mean the church must or should use it.

We need to grapple with the inherent dangers posed by livestream technology, ensure that biblical and theological values are in the driver's seat, and then prayerfully and carefully consider how best we can glorify our King in our particular ecclesiastical contexts. That is precisely what this book sets out to help you do.

---

16. I'm indebted to the insight of T. David Gordon for this distinction between commanding and commending (*Choose Better: Five Biblical Models for Making Ethical Decisions* [Phillipsburg, NJ: Presbyterian & Reformed, 2024], 5). We will come back to his five ethical models in the conclusion with particular application to livestream.

CHAPTER 1

BEWARE OF AN ILLEGITIMATE AUTHORITY

DIGITAL TECHNOLOGY is an astounding thing. With the click of a button, the walls of my study begin to reverberate with sounds that in previous eras could only be heard by going to a concert hall. A good book with a hot cup of coffee is raised to a whole new level of delight when enjoyed against the backdrop of the symphonic harmony of a well-trained orchestra.

How do all of those musicians join their stringed instruments together to create such audible ecstasy? The answer is simple—authority.

What we can easily forget when listening to an orchestra on Spotify is that leading this body of talented musicians is a little man dancing wildly with a wand in his hand. He is the conductor, and he is in charge. Every member of the orchestra follows his lead, and as they do, they are able to reach heights of musical brilliance which they never could on their own. The political scientist David Koyzis, defending the necessity and legitimacy of authority in the various spheres of human society, makes this very point, stating, "One trombonist cannot play Beethoven's *Eroica*, but in concert an entire orchestra can do so under the baton of a conductor. In this respect, authority has expanded the range of the trombonist's freedom."[1] In a day when authority is perceived as inherently oppressive, the orchestra serves as a beautiful reminder that following legitimate authority is actually liberating, enabling us to reach our full human potential in ways we couldn't without it.

The Cambridge Dictionary defines authority as "the power to control or demand obedience from others."[2] Merriam-Webster defines it as "power to command thought, opinion, or behavior."[3] According to the leading English dictionaries, authority is a commanding power.

The Scriptures are in hearty agreement from the start. On the very first page we meet a God who exercises infinite power to command the universe into orderly existence out of nothing (Gen. 1:3–31). No sooner does He speak creation into existence do we then find Him wielding commanding power to prohibit the first humans from eating the fruit of a tree upon the pain of death (Gen. 2:15–17). Sound oppressive? That is what the serpent told Adam and Eve, portraying God as a tyrant who exercises commanding power to harass humanity (Gen. 3:4–5). But nothing could have been further from the truth! As the first priests in the first earthly temple, Adam and Eve would

1. David T. Koyzis, "Why We Need Authority," *Comment*, September 9, 2011.

2. https://dictionary.cambridge.org/us/dictionary/english/authority.

3. https://www.merriam-webster.com/dictionary/authority.

know joyous liberation only as they served God in submission to His authoritative word.[4] When they heeded the devil's lie instead, they didn't cease yielding to an external authority. They simply traded God's rightful authority for Satan's wrongful authority, and in so doing they forfeited life and blessedness and were exiled from God's garden-temple (Gen. 3:22–24).

The history of redemption is the story of God delivering a people from Satan's oppression and restoring them to life-giving, soul-satisfying relationship with Himself. The Bible calls that divine-human relationship a covenant. Similar to the covenant of marriage, God's covenant with His people is a loving communion bond wherein both parties give themselves exclusively to the other. But unlike marriage, which is a bond between two image-bearing creatures who are ontological equals, God's relationship with us is not a bond between equals. There is an existential gulf of infinite proportions between the parties of this covenant! Simply put, God is Creator, and we are creatures. That means at every point in the relation, He is authoritative Lord, and we are servants.[5]

To return to our analogy, God is the Conductor, and if we decide to snub His authority and play our cello however we want, He has the right to kick us out of His orchestra. Such was the story of Adam, and such was the story of Israel (functioning as a corporate Adam of sorts). It's why the Son of God came as the second Adam—to rescue God-hating rebels from their

---

4. G. K. Beale, *The Temple and the Church's Mission: A Biblical Theology of the Dwelling Place of God*, New Studies in Biblical Theology 17, ed. D. A. Carson (Downers Grove, IL: InterVarsity, 2004), 66–87.

5. To use the language of ancient Near Eastern covenant treaties, God is the Suzerain and we are His vassals.

covenant-breaking and to transform them into God-loving servants who follow the divine Conductor wheresoever He leads.

The God-Directed Music of the Redeemed

What is the music God leads His redeemed people to play through Christ's mediation? Worship.[6] That is, after all, what priests do. God's ultimate aim in creating us in Adam and recreating us in Christ is His worship. Even my four-year-old son can tell you that his chief purpose in life "is to glorify God and to enjoy Him forever" (Westminster Shorter Catechism 1). We are called to serve God with hearts that are seeing and savoring Him.

John Calvin, laboring for reformation in the European church, argued that "the whole force of Christianity" could be comprehended in two things: "that people may know (1) how God is rightly worshiped and (2) whence they must seek salvation for themselves."[7] Did you hear that? According to Calvin, God's worship rightly carried out by God's saved people is a comprehensive description of biblical religion. Notice that he puts knowing how to worship before knowing how to be saved. Calvin understood what we are quick to forget—God's glory is the chief end of our existence and our exodus. Worship is the music of the redeemed, and according to Calvin, it is of supreme importance how that music is played. Given the nature

---

6. By this, I don't mean to say that worship is only through song as is often assumed today. Singing to God via music is only one of a number of ways God's people worship Him. Here I am simply continuing to draw upon the illustration of an orchestra.

7. John Calvin, *The Necessity of Reforming the Church with a Reply to Cardinal Sadoleto,* trans. Casey Carmichael (1554; Sanford, FL: Reformation Trust Publishing, 2020), 6.

of the covenant, we don't dare serve Him in any way we please. He is Lord, and He calls the shots.

As we consider the use of livestream technology in the church's worship, we must begin with God's authority as Creator and covenant Lord to regulate how we affectionately exalt Him.[8] We can't begin with the shut-ins unable to attend worship or the neighbors unwilling to step into our church buildings. We must begin with God and what historically came to be called *the regulative principle of worship.*

The opening paragraph of chapter 21 of the Westminster Confession of Faith succinctly articulates this foundational truth: "The light of nature showeth that there is a God, who hath lordship and sovereignty over all, is good, and doth good unto all, and is therefore to be feared, loved, praised, called upon, trusted in, and served, with all the heart, and with all the soul, and with all the might. But the acceptable way of worshiping the true God is instituted by himself, and so limited by his own revealed will, that he may not be worshiped according to the imaginations and devices of men, or the suggestions of Satan, under any visible representation, or any other way not prescribed in the Holy Scripture."

As image-bearing creatures living in a world that everywhere reveals the glory of God, humans have an innate knowledge of their Creator and the worship He is due. But even before the fall, Adam was not free to serve God however he pleased. He was covenantally bound to the revelation of God's will in His spoken word. It was God's command that regulated

---

8. I define worship as the affectionate, whole-person exaltation of our infinitely valuable God. By affectionate, I'm not referring to a flippant emotion, but to the Spirit-wrought, reverent, happy love toward God that fuels all true worship.

how Adam was to carry out His priestly duties in the Edenic temple (Gen. 1:28; 2:15).

After the fall, the need of this word became all the more necessary given humanity's idolatrous propensity, not only to worship other gods, but to worship the true God in the wrong way. As Thomas Boston laments, "There is a special proneness in the [fallen] nature of man to corrupt the worship and ordinances of God."[9] Left to our own devices, we don't know what notes to play or when to play them, and in our sin, we find a sick delight in composing our own insufferable noise against the divine Composer's will. The result? Covenant curse.

But the good news of the gospel is that Christ suffers our curse, reconciling us to God and delivering us from our idolatrous ways (1 Thess. 1:9–10). In so doing, He restores us to the priestly vocation of worship (Exod. 19:5–6; 1 Pet. 2:9). Given that the ultimate aim of our existence is worship, there is arguably no greater responsibility of the redeemed than to ensure that God's worship is faithfully carried out. It is a responsibility that requires constant vigilance. For, in the words of John Owen, "there are many mistakes, even amongst the saints themselves in their apprehensions in and about the worship of God."[10] Until we reach glory, we are ever prone to worshiping God in ways that are unbefitting of His glory and that ultimately undermine His sovereign authority. So when something like livestream comes along, the church has to think

---

9. Thomas Boston, *An Illustration of the Doctrines of the Christian Religion*, in *The Complete Works of Thomas Boston* (Stoke-on-Trent, UK: Tentmaker, 2002), 2:129.

10. John Owen, "The Nature and Beauty of Gospel Worship," in *The Works of John Owen*, ed. William H. Goold (Edinburgh: Banner of Truth, 1965), 9:71.

very carefully before introducing it into God's worship. What may be appropriate for a news broadcast could be destructively inappropriate for the worship of God's people.

Our forefathers didn't develop the regulative principle out of thin air, nor did they derive it solely from the Creator-creature distinction. They rightly grasped that how we worship was of such significance in the eyes of God that He wrote it upon Adam's heart in creation (Rom. 2:14–15) and later etched it into Israel's constitution on stone tablets at Sinai (Exod. 20:1–17). Where exactly? In the Second Commandment: "You shall not make for yourself a carved image, or any likeness of anything that is in heaven above, or that is in the earth beneath, or that is in the water under the earth." (Ex. 20:4).

A Surprising Reference to the Decalogue

In the 1980s Neil Postman wrote a prophetic work titled *Amusing Ourselves to Death: Public Discourse in the Age of Show Business*. In it, he lamented how the modern entertainment industry had transformed the way western society communicated and reasoned. It was not the least bit surprising to find Postman quoting Marshall McLuhan's well-known adage "the medium is the message."[11] As we argued in the introduction, technological mediums have a message of their own, regardless of their content. Furthermore, as Postman draws out in his book, those mediums determine the content being communicated and thereby exercise a tremendous power over the way a culture thinks. All of that we would expect when reading McLuhan and Postman. But what completely shocked me when I read *Amusing Ourselves to Death* was that immediately after

---

11. Neil Postman, *Amusing Ourselves to Death: Public Discourage in the Age of Show Business* (New York: Penguin, 1985), 8.

quoting McLuhan, Postman argues that all of this can be found in a much more ancient source than McLuhan. His words are worth quoting at length:

> In studying the Bible as a young man, I found intimations of the idea that forms of media favor particular kinds of content and therefore are capable of taking command of a culture. I refer specifically to the Decalogue, the Second Commandment of which prohibits the Israelites from making concrete images....I wondered then, as so many others have, as to why the God of these people would have included instructions on how they were to symbolize, or not symbolize, their experience. It is a strange injunction to include as part of an ethical system *unless its author assumed a connection between forms of human communication and the quality of a culture.* We may hazard a guess that people who are being asked to embrace an abstract, universal deity would be rendered unfit to do so by the habit of drawing pictures or making statues or depicting their ideas in any concrete, iconographic forms. The God of the Jews was to exist in the Word and through the Word, an unprecedented conception requiring the highest order of abstract thinking.[12]

Postman, who was a secular Jew, understood that Israel's mode of worship was no indifferent matter. To worship the true God through a medium unbefitting of His glory would ultimately lead to worshiping something other than God. Embedded in the Decalogue itself was the principle that the tools we use to communicate with and about God are not neutral. They themselves communicate a message and control the content that is able to be communicated.[13]

---

12. Postman, *Amusing Ourselves to Death*, 9, emphasis original.

13. If I understand Postman rightly, I tend to agree with him that it is better to speak of mediums as metaphors rather than messages (*Amus-*

In the Second Commandment, God prohibits His people from worshiping Him through the medium of images because that medium fundamentally denies His independent, infinite, incorporeal glory as God. Image worship entails the creature fashioning a likeness of the Creator. This is altogether inappropriate. The independent Creator has the right to fashion a creature as His image (Gen. 1:27), but the creature has no authority to fashion an image of the Creator. Furthermore, image worship entails the Creator being fashioned in the likeness of the creature. That is why God forbids us from making a likeness of Him according to "anything that is in heaven above, or that is on the earth beneath, or that is in the water under the earth." To craft an image necessarily denotes losing the Creator in the creation, fashioning Him after the likeness of a beast or a man. In other words, the medium has a message of its own— the Creator is fashioned by the creature after the likeness of the creature. Regardless of what the image is, we are already being told what God is like. The problem is, He is not like that at all.

The idea that the medium of worship doesn't matter, so long as we worship the true God, is, to borrow and modify the language of McLuhan quoted in the introduction, *the numb stance of the theological idiot*. It doesn't just matter who we worship; it also matters how we worship. For, as Ligon Duncan wisely contends, "how we worship determines who we worship."[14]

The Full Scope of the Second Commandment

But did you notice how the Westminster divines didn't just prohibit images in God's worship, but went so far as to prohibit

---

ing Ourselves to Death, 10–15). Mediums don't communicate, "Reality is…" Instead, they communicate, "Reality is like…"

14. J. Ligon Duncan III, "Does God Care How We Worship?" in *Give*

"any other way not prescribed in the Holy Scripture"? Where did they get that? Ultimately, from their understanding of the entire Bible! But they also perceived that the Decalogue, as a summary of God's revealed will, utilizes synecdoche through-out.[15] Synecdoche is a figure of speech wherein a primary part of something represents the whole.

One of the elders I serve alongside of drives a 2007 Honda Fit. When I first visited the church the summer after graduating seminary, Dan drove me all over town in his little red car. I distinctly remember tightly gripping the door as he zipped around the sharp curves of Lookout Mountain. Imagine that, after surviving the trip, I got out and exclaimed, "Those are some sweet wheels!" Dan may have wondered if I was being sarcastic, but he wouldn't have wondered why I was praising his rubber Michelins. It would be obvious that I was using synecdoche, speaking of the wheels as representative of the entire car.

In the ancient Near East, man-made idols in the forms of statues and images were everywhere. So, God gave Israel the chief example of an idolatrous medium of worship to represent a broader category of all unlawful means of exalting Him. It is a warning that we must take great care how we approach God.

Sometimes the Westminster divines (and the Puritans more broadly) are portrayed as if they invented this interpretive

---

Praise to God: A Vision for Reforming Worship, eds. Philip Graham Ryken, Derek W. H. Thomas, and J. Ligon Duncan III (Philipsburg, NJ: Presbyterian & Reformed, 2003), 33.

15. They explicitly spell this out in their interpretive principles for the Decalogue in Larger Catechism 99. One of those eight principles is synecdoche: "That under one sin or duty, all of the same kind are forbidden or commanded; together with all the causes, means, occasions, and appearances thereof, and provocations thereunto."

principle in order to stretch the meaning of the Decalogue to support their kill-joy, constrictive, legalistic model of religion. In reality, they got this principle from the way God Himself interprets and applies the Decalogue.

The book of Deuteronomy preserves Moses's death-bed preaching to the Israelites as they prepare to conquer the Eden-like garden of Canaan. Moses's message to Israel was God's covenant, and in chapters 6–26 he expounds and applies the Decalogue (as the charter of the covenant) to Israel's unique existence in the land of Canaan.[16] After preaching the First Commandment in chapters 6–11, calling Israel to exclusive, reverent, joyful devotion to God, Moses turns to proclaim the Second Commandment in chapter 12. We'll look more fully at his exposition in the next chapter, but his final words summarize the regulative principle well: "Everything that I command you, you shall be careful to do. You shall not add to it or take from it" (Deut. 12:32). In the context, Moses is speaking of the formal worship that Israel is to render to God in Canaan. They are to worship according to God's revealed will alone, neither adding nor subtracting anything. As the Shorter Catechism positively states, "The second commandment requireth the receiving, observing, and keeping pure and entire, all such religious worship and ordinances as God hath appointed in his Word" (a. 50).

Here is the heart of the regulative principle—the Lord whom we worship is the Lord of His worship. It is necessary for us to take time to establish this because unfortunately it is a much overlooked fact and one that rarely, if ever, comes up

---

16. Many scholars have made this point. See, for example, Allan Harman, *Deuteronomy: The Commands of a Covenant God,* Focus on the Bible Commentary Series (Fearn, Ross-shire: Christian Focus, 2001), 12–14.

when discussing livestream in worship. To make matters worse, disregarding this principle can have deadly consequences.

Strange Fire and the Fire of God's Wrath

Many of us begin the new year resolved to read through our Bibles from Genesis to Revelation. We get off to a good start until we hit the middle of Exodus and begin to drown in the detailed regulations for constructing the tabernacle. The more disciplined among us persevere and reach Leviticus only to meet nine consecutive chapters of extremely precise blueprints for how God's priests were to approach Him. Why would God include these extensive ceremonial laws in His word, and why should we bother to read them? One reason is that God is making crystal clear that when it comes to His worship, nothing is left to the will of man. Speaking of tabernacle worship, Owen writes, "There was not the least part of the fabric wherein his worship was celebrated, nor any ornament of it,—not one rite or ceremony that did attend it,—but it was all of it wholly of God's own designation and command."[17] By the time we reach the middle of Leviticus, it is obvious to any attentive reader that God is the authoritative Lord of His worship.

We might breathe a sigh of relief when we finally reach historical narrative in Leviticus 10 after all the detailed laws and ceremonies. But the relief only lasts for one verse before the priestly sons of Aaron are burned alive by the fiery glory of the God they sought to worship. God consumed them in His wrath as they attended to their duties in the tabernacle. Why? It was not because they sought to worship another god, violating the First Commandment. The reason was that they "offered *unauthorized* fire before the LORD, *which he had not com-*

---

17. Owen, "The Nature and Beauty of Gospel Worship," 9:79.

manded them" (Lev. 10:1, emphasis mine). They thought they could worship God by a means not authorized in His word, and they didn't get a chance to think again. Nadab and Abihu would have done well to study the Second Commandment, especially the reason God attached to it: "for I the LORD your God am a jealous God…" (Exod. 20:5). God burns with a holy zeal for His worship and will not tolerate what our forefathers called "will-worship," that is, worship offered up according to our own wills and whims. The Puritan Jeremiah Burroughs published a whole book on this tragic incident in the tabernacle, stating, "We must all be willing worshippers, but not will-worshippers. We must come freely to worship God, but we must not worship God according to our own wills."[18] Do so, and you won't just forfeit a true knowledge of God, but you might just forfeit your life.

The elaborate ceremonies of the old covenant have passed away with the coming of the Son of God in human flesh as the ultimate tabernacle, priest, and sacrifice (Jn. 1:14; 2:18–22; Heb. 9:11–14). Because of Christ's finished work, new covenant worship is marked by fewer ordinances "with more simplicity, and less outward glory" (Westminster Confession 7.6). But it is not therefore the case that God has left it up to the whims of His people to determine how they might approach Him. If anything, our greater new covenant privileges call us to a heightened responsibility to jealously guard the worship of our jealous God. That is the logic of the writer to the Hebrews when, after waxing eloquent about the heavenly superiority of new covenant worship, he commands us to "offer to God acceptable worship, with reverence and awe, for our God

---

18. Jeremiah Burroughs, *Gospel Worship: Worship Worthy of God*, ed. Don Kistler (Orlando, FL: Soli Deo Gloria, 1990), 11.

is a consuming fire" (Heb. 12:28–29). If you are thinking of Nadab and Abihu, you are thinking rightly. The fiery God who consumed them hasn't changed. We must approach Him in an acceptable way on His terms alone.

Elements, Forms, and Circumstances

The regulative principle, which is in fact a biblical principle, must be at the foundation of all considerations about worship, including the technology we use in it. But before we apply the regulative principle to the use of livestream, we need to consider an important distinction. While God regulates the elements of worship by the specific commands of His word, He regulates the forms and circumstances of worship by the general principles of His word.

The *elements* of worship refer to the means we utilize to approach and exalt God. The only elements that are acceptable in God's worship are those which He has expressly commanded. Anything beyond that is like the strange fire of Nadab and Abihu. The authorized elements of new covenant worship include reading the word (1 Tim. 4:13), preaching the word (2 Tim. 4:2), praying the word (1 Tim. 2:1), singing the word (Eph. 5:19), and partaking of the sacraments which signify and seal the word (Matt. 28:19; 1 Cor. 11:23–29).[19] These are the new covenant elements of worship regulated by and saturated with God's self-revelation in the Scriptures. They are how we affectionately exalt God as His new covenant priests in Jesus Christ.

Then there are *forms* which address how we carry out the elements. For example, the sacrament of the Lord's Supper is an element, but it can be partaken of via different forms. One church might drink from a single silver goblet, while another

---

19. Some also include the giving of tithes and offerings (1 Cor. 16:2).

might drink from individual plastic cups. One church might offer both wine and grape juice, while another might offer only wine or grape juice. It is the same element being engaged through different forms, and each church chooses its particular form upon the basis of general principles from God's word. The one serves wine and grape juice in individual cups upon the basis of the biblical principles of love for neighbor and liberty of conscience. The other serves only wine or grape juice from a single cup upon the basis of the biblical principle of the unity of the body of Christ. But both are regulated by God's word. I might prefer cold brew coffee and donuts over wine and bread, but I don't have the right to introduce those forms at the Table. To do so would be to lose the sacrament altogether, introducing an unauthorized element into worship (namely, coffee and donut time!). Certain forms are inherently idolatrous, and all forms must be informed by God's word. There is, however, significant room for variation in the forms by which we carry out every one of the God-prescribed elements of new covenant worship.

Finally, there are *circumstances*. These include things like what time a church meets for worship on the Lord's Day and how long they meet. God has prescribed a day—the first day of the week—on which His people are to gather to worship Him, but He hasn't spelled out what time or for how long.[20] One church might meet for worship on Sunday at 10:00am, enjoy fellowship over lunch, and then meet again at 1:30pm, because the congregation is too spread out geographically and it

---

20. For a brief defense of the Lord's Day as the Christian Sabbath see D. G. Hart and John R. Muether, *With Reverence and Awe: Returning to the Basics of Reformed Worship* (Phillipsburg, NJ: Presbyterian and Reformed, 2002), 61–73.

is impractical for most members to drive back and forth twice on a Sunday. Another church might rent their meeting space, requiring them to worship at 8:00am before the building is occupied by others. The apostle Paul preached so long and late into the night on the first day of the week that a little boy fell asleep, fell out a second-story window, and died (Acts 20:7–9). But as certain members of my church like to remind me, that is not an apostolic injunction for me to go and do likewise! Every pastor must assess his own gifts and the constitution of his congregation to determine the ideal length of his sermons. These circumstantial matters have direct bearing upon the worship of God, but they are not regulated by explicit commands from God. Our Lord wills for them "to be ordered by the light of nature, and Christian prudence, according to the general rules of the Word, which are always to be observed" (Westminster Confession of Faith 1.6).

So How Do We Classify Livestream Technology?

These categories are helpful as we think through the relationship of livestream to God's worship. Should this digital technology be classified as an element, a form, or a circumstance?

What is obvious is that livestream, when used in worship, is not an element. Unlike skits, dance routines, and movies clips which are sometimes brought into a church's worship as unauthorized elements, livestream is not another thing we *do* in worship. It is rather the way we do what we do in worship. This is an important point because it means that whatever conclusions we might draw about livestreaming technology in public worship, it is not directly in the same category as the old covenant priests crafting a statue or offering up an unprescribed sacrifice.[21] Such things are elements, and preserving the proper elements in God's worship is worth living and dying for. But take careful note that livestream (at least in the way it is used in the church today) is not an element. Why does that matter? Because, as Jonathan Cruse laments, "Sadly, many churches spend a lot of time arguing about circumstances as though they held the authoritative place of elements."[22] How many people leave churches because the preacher is a bit too long-winded or the music doesn't quite scratch them where they itch? Elements must be the big thing, and unless a form or circumstance is being introduced that destroys or significantly damages the element, we ought to take great care to show deference to fellow brothers and sisters in the Lord who differ from us.

If livestream is not an element, then what is it? One needs to only consider the pandemic to realize that circumstances are often what lead someone to use this technology in worship. But livestream is not itself a circumstantial matter. When utilized to participate in the elements of public worship, it is best classified as a form. For livestream actually determines how we engage with God through the public means of grace.

Remember that when it comes to forms, it is not an anything-goes free-for-all. What exactly, then, are we looking for in a form? Ligon Duncan perceptively notes that "the *forms* in which [the] elements are performed must not be inimical to the nature or content of the element or draw attention away

---

21. Though it could possibly be indirectly in that category as certain forms are so inimical to the element that they actually introduce a new element (e.g., coffee and donuts at the Table).

22. Jonathan Landry Cruse, *What Happens When We Worship* (Grand Rapids: Reformation Heritage Books, 2020), 65.

from the substance and goal of worship."[23] Later on in the same article he writes,

> The Reformed tradition has not been concerned with forms and circumstances so much for their own sake as much as for the sake of the elements and substance of worship and for the sake of the object and aim of worship. The Reformers....understood that the liturgy, media, instruments, and vehicles of worship are never neutral, and so exceeding care must be given to the 'law of unintended consequences.' Often the medium overwhelms and changes the message.[24]

You might want to go back and read those words again. The vehicles and mediums by which we carry out the elements are not neutral, including the medium of livestream. We need to exercise "exceeding care" to ensure that a medium such as this is not undermining the substance, elements, or aim of worship. If at any point that is the case, then the church has inevitably drifted into idolatrous will-worship.

We will spend the rest of the book grappling with whether the particular medium of livestream is lawful or not, and in the end, we might not be in perfect agreement. The legitimacy or non-legitimacy of many forms are not as straightforward as Starbucks at the Table. Livestream is a particularly knotty one, but as Christians we cannot afford to accept this (or any) liturgical medium without thinking deeply about it and its potential consequences. For there is nothing more important than the

---

23. J Ligon Duncan III, "Foundations for Biblically Directed Worship," in *Give Praise to God*, 55-56.

24. Duncan, "Foundations for Biblically Directed Worship," 64.

public worship of God, and the forms we use have the power to shape our worship profoundly. We must ask three questions:

1. Does livestream undermine the substance of public worship? We will explore that in chapter 2 as we think about the necessarily public nature of public worship.
2. Does livestream undermine the elements of public worship? We will explore that in chapter 3 as we think about the importance of bodily, physical presence in God's worship.
3. Does livestream undermine the aim of public worship? We will explore that in chapter 4 as we think about the ways in which using this technological tool in the church might cater to self-serving idolatry, making us apathetic to the weighty glory of the Triune God.

Failure to face these questions leaves us vulnerable to an illegitimate authority, threatening to keep us from playing the music of the redeemed at the lead of our heavenly Conductor. It could cost us both our lives and our eternities.

Statism and the Church's Worship

Before moving on, there is one other matter, pertaining to authority and the worship of God, that we must consider. In the midst of the Covid-19 pandemic, many civil leaders took it upon themselves to shut down society as we know it, including the church. Only that which the magistrate labeled an "essential service," such as grocery stores and hospitals, were allowed to remain open. According to many, the embodied gathering of the church to serve God in worship and one another in fellowship, was non-essential. Thus, church closures were mandated. After those mandates were lifted, the civil government in many places issued authoritative regulations for the church's worship, requiring masks and social distancing, and even prohibiting certain elements of worship, like singing.

It goes without saying that the civil authorities were dead wrong when they told us that public worship was not an essential service. Arguably, it is the most essential service—the service of all services! Exploring that, however, would take us beyond the bounds of our purposes here. What is pertinent for us is this question: Does the civil magistrate have authority to prescribe if, when, or how the church can worship God?

If 2020 revealed anything, it was that the church in the West hadn't thought much about that question. It also revealed just how many Westerners, Christian and non-Christian alike, had come to accept the all-encompassing authority of state. At certain points in history, the predominant view of society has been that the church had authority over the affairs of the state. In our day, the prevailing view is that the state has authority to meddle in the affairs of the church. This means that, if in the interest of public health, the state tells us we cannot gather for worship, or that we cannot sing in worship, or that we cannot embrace one another in fellowship, it seems reasonable and right to submit. After all, that is what Romans 13 teaches, isn't it?

Not exactly. Romans 13:1–7 teaches us that God possesses absolute authority (v. 1: "there is no authority except from God") and that God delegates some of that authority to those He appoints to lead in the civil sphere. The magistrate doesn't possess absolute authority; he possesses a limited authority (v. 4: "the sword") for a particular purpose (v. 4: "he is a servant of God, an avenger who carries out God's wrath on the wrongdoer"). God appoints civil leaders and endows them with authority to restrain evil and to promote moral order in society, lest the human race self-destruct in murderous chaos.

As citizens of the state, individual Christians ought to support and submit to their civil leaders, respecting their God-given authority. But they do so understanding the limitations of that authority. Herman Ridderbos, commenting on Romans 13, explains, "All this means—in the nature of the case—not that the church is here summoned to a 'blind' obedience to government and that Paul declares every existing governmental power inviolable and sacrosanct."[25] We are to submit to the civil government when they wield legitimate authority. But when they pry into matters not delegated to them by God, they are exercising an illegitimate authority, and we ought not to blindly adhere to it.

The Church's Divine Right

As the source of all authority, God has designed various spheres in human society wherein He appoints leaders who are delegated a limited authority for particular ends.[26] The church and the state are two of those spheres. Speaking of them, James Bannerman writes,

> They have a separate jurisdiction; they have separate organs and office-bearers to exercise it. There is a magistracy that appertains to the state,—the appointment and ordinance of God to exercise the functions which God has intrusted to the state. There is a different magistracy that appertains to the Church,—the

25. Herman Ridderbos, *Paul: An Outline of His Theology*, trans. John Richard De Witt (Grand Rapids: Eerdmans, 1975), 324.

26. For a helpful introduction to sphere sovereignty see Joseph Boot, *Ruler of Kings: Toward a Christian Vision of Government* (London: Wilberforce, 2022), 171–203.

appointment and ordinance of Christ to discharge the duties which Christ has intrusted to the Church. The two are wholly apart from each other, and cannot interchange office or authority or duty.[27]

It is not within the state's jurisdiction to determine matters pertaining to the church's doctrine or worship. When the state authoritatively meddles in ecclesiastical affairs and the church unquestioningly submits, God's people are subjected to an illegitimate authority. Listen to how Herman Bavinck puts it: "Those who...shrink this power of the church, limit it, and assign it to the civil government diminish the honor of Christ and fail to do justice to the rights and freedoms granted to the church."[28]

Christ, as authoritative Lord, endows authority to the leaders of the church to oversee her doctrine, worship, and discipline. While the state is given an authority symbolized by the sword (Rom. 13:4), the church is given an authority symbolized by the keys: "I will give you the keys of the kingdom of heaven, and whatever you bind on earth shall be bound in heaven, and whatever you loose on earth shall be loosed in heaven" (Matt. 16:19). Jesus, speaking to the initial apostolic leaders of the church, likens them to Eliakim, the chief administrator of David's kingdom, entrusted with the keys to the royal palace (Isa. 22:22). In ancient times, it was common for such keys to be worn around the shoulder as a sign of authority. Just as Eliakim was the steward of the keys to David's royal house, so too the

---

27. James Bannerman, *The Church of Christ: A Treatise on the Nature, Powers, Ordinances, Discipline, and Government of the Christian Church* (Edinburgh: Banner of Truth, 2015), 108.

28. Herman Bavinck, *Reformed Dogmatics*, ed. John Bolt, trans. John Vriend (Grand Rapids: Baker Academic, 2008), 4:414.

apostles, represented here by Peter, were appointed as stewards of Christ's palace, the church. They were entrusted with the royal keys to open and close the kingdom through preaching, administrating the sacraments, and exercising discipline.[29] In the post-apostolic era, that authority has been entrusted to the elders of the church (1 Pet. 5:1–4; Acts 20:28).

It is the elders who call the local church to worship. It is the elders who are responsible to ensure the elements of worship are being carried out through forms and circumstances that honor Christ and bless His people. It is the elders who have the right to cancel a worship service in the face of a natural disaster or emergency (there are legitimate reasons to cancel Lord's Day worship).

The state may appeal to the church to consider temporarily closing its doors. The state may propose recommendations for how the church worships in a time of plague. But when the state prohibits public worship or mandates how the church worships under threats of fines and jailtime, it is exercising an authority not given to it by God.[30] Bannerman contends,

> The Church has a right, from her Divine Head, to the full possession and free use of all the powers and prerogatives which

---

29. For a helpful exposition of the keys, see Guy Prentiss Waters, *How Jesus Runs the Church* (Phillipsburg, NJ: Presbyterian & Reformed, 2011), 29–54.

30. I've heard people liken the Covid lockdowns and mandates to building codes and maximum occupancy requirements. The state tells my church that we can only have 200 people in our meeting place at any given time. Is there a problem with that? Not necessarily, though the civil authorities can easily get out of hand in their regulation of private property. But the regulation of a physical building in the service of public safety is vastly different from the regulation of the spiritual building of the church in the service of public safety. I don't object to

He has vested in her, without interference or obstruction of any kind from the civil magistrate. In preaching the truth according to Christ's Word, in administering ordinances according to His appointment, in exercising authority and discipline in conformity with His gift and injunction, the Church must be free to judge and act for herself according to the law of Scripture, without responsibility to or interference from the state.[31]

We do well to write those words upon the doorposts of our hearts. For while the Covid-19 lockdowns may be a relic of the past, it won't be the last time civil authorities seek to tell the church in the West if, when, or how she may worship.

All of this might seem like an extended rabbit trail, but I assure you it is not. How many churches accepted livestream, not because the elders had carefully and prayerfully determined it to be a fitting form in which to carry out the elements of public worship, but because they believed they were duty-bound to submit to the state's prohibition from gathering in flesh-and-blood physicality and so resorted to livestreaming worship as the next-best thing?

Even if livestream doesn't violate God's will for His worship, to accept livestream in God's worship because of the unlawful demands of the state does violate God's revealed will for His worship. For it is the elders of the local church, filled with the Spirit of Christ and submitted to the word of Christ, who have the authority to determine if the church assembles and to

---

the state telling us how many people we can have in our physical meeting place. Even if I wish the number were different, I joyfully submit to their orders. But I do object to the state telling us that we can't gather as the spiritual temple of the living God. They have no right to do so.

31. Bannerman, *The Church of Christ*, 118–19.

receive or reject livestream or any other technological novelty in the public assembly of His people. While Christ conducts the universal church's worship from heaven, He appoints elders to represent Him in the local church's worship on earth, leading the church in the harmonious melody of the redeemed. Nothing poses a greater threat to that soul-liberating, symphonic sound than an illegitimate authority.

CHAPTER 2

BEWARE OF AN
INESSENTIAL ASSEMBLY

THE COVID-19 LOCKDOWNS that resulted in the widespread normalization of livestreaming worship services also resulted in the realization among tech companies that there was major money to be had in making religious institutions dependent upon their platforms. In the summer of 2021, Elizabeth Dias wrote an article for *The New York Times* titled "Facebook's Next Target: Religious Experience." Dias chronicled how the pastor of a well-known church "sought advice on how to build a church in a pandemic," not from the Scriptures, church history, or wise fellow pastors, but from Facebook. He began meeting with Facebook staff weekly to explore "what the church would look like on Facebook and what apps they might create for financial giving, video capability or livestreaming."[1]

---

1. Elizabeth Dias, "Facebook's Next Target: Religious Experience," *The New York Times,* July 25, 2021.

Let that sink in for a moment. This pastor sought advice from Facebook on how to *build a church,* even signing a non-disclosure agreement with the company.

Four years earlier, the founder and CEO of Facebook, Mark Zuckerberg, had hailed his social media platform, with its billions of users, as a replacement for the church.[2] But after thousands upon thousands of local congregations began streaming their services on Facebook, Zuckerberg and his team realized there was potential, not to replace the church, but to own the church, along with the entire realm of religious experience. Dias explains, "Now, after the coronavirus pandemic pushed religious groups to explore new ways to operate, Facebook sees even greater strategic opportunity to draw highly engaged users onto its platform. The company aims to become the virtual home for religious community, and wants churches, mosques, synagogues and others to embed their religious life into its platform, from hosting worship services and socializing more casually to soliciting money."[3] Those two sentences are quite revealing. Facebook sees the livestream revolution as a "strategic opportunity" to grow its massive body of "highly engaged users." The church is a means to that end, and if that end is to be realized, Facebook must be successful in leading churches "to embed their religious life into its platform."

"I Will Build My Church"

To seek counsel from a transhumanistic, anti-Christian tech company as to how to build the church evidences that you don't understand the first principle upon which the church's

---

2. https://www.cnbc.com/2017/06/26/mark-zuckerberg-compares-facebook-to-church-little-league.html.

3. Dias, "Facebook's Next Target."

existence hinges. That principle is articulated in four Greek words from the lips of our Lord translated into five English words: "I will build my church" (Matt. 16:19). When we begin to look to human wisdom and innovation to build the church instead of Christ, the building we construct might be branded a church, but in reality it is just another iteration of the tower of Babel (Gen. 11:1–9).

The devil loves few things more than Christian leaders who put their trust in mere men to construct the church rather than the God-man. For the result will be a counterfeit church which is fit to lull many into a spiritual slumber, all in the name of serving Christ. It is one of Satan's most brilliant strategies, but it will ultimately fail. For Christ, as the Builder of the church, has promised that "the gates of hell shall not prevail against it" (Matt. 16:19). He will see to it that in every generation there is a multitude of faithful worshipers who do not bow the knee to the cultural Baals (1 Kgs. 19:18). We need not fear when we read headlines about the devilish intentions of Facebook (or any other tech company) for the church. Christ doesn't assure the apostles that He is going to try His best to build His church. He *will* do it—with or without us!

The English word *church* comes from the Greek word *ekklesia*. Francis Turretin fleshes out its etymology for us: "It designates both a separation by the force of the preposition *ek* and a collection and congregation from the emphasis of the verb *kaleo*, so that it is a society of men called out of some place or state and congregated into an assembly."[4] Matthew is the only gospel writer to use the term (16:19; 18:17). His

---

4. Francis Turretin, *Institutes of Elenctic Theology*, trans. George Musgrave Giger, ed. James T. Dennison Jr. (Phillipsburg, NJ: Presbyterian & Reformed, 1997), 3:6.

predominantly Jewish readers would have grasped intuitively the significance of the word, for it is used all throughout the Greek Old Testament to refer to Israel as those called out of an idolatrous world to assemble in God's holy presence. When Jesus declares, "I will build my church," He is literally saying, "I will build my assembly." Explaining the term *ekklesia* in its biblical significance, John Murray writes, "It is not so much the called-out ones as the called-together ones, going back to the Old Testament assembly before the glory of God's self-manifestation at Sinai."[5] The exodus, as the ultimate picture of God's salvation under the old covenant, reached its culmination in God's liberated people congregating at the foot of the mount to meet with Him through the mediation of Moses.[6] Israel was redeemed together to assemble together. That initial corporate gathering at Sinai became a paradigm for Israel's worship from that point forward.

God's Chosen Place Under the Old Covenant

Where did God call His people to assemble after the Sinai event? Moses describes it as "the place that the LORD your God will choose out of all your tribes to put his name and make his habitation there" (Deut. 12:5). As God's prophet prepared the people to conquer Canaan, he expounded and applied the Second Commandment in Deuteronomy 12, calling Israel to worship according to God's revealed will alone. While much of his exposition focuses upon the elements of worship (i.e., the old covenant offerings prescribed by God), it equally focuses

---

5. John Murray, "Worship," in *Collected Writings of John Murray* (Edinburgh: Banner of Truth, 1977), 1:166.

6. Edmund P. Clowney, *The Church,* Contours of Christian Theology, ed. Gerald Bray (Downers Grove, IL: InterVarsity Press, 1995), 30.

upon the place of worship. God didn't just regulate *what* Israel did in worship, He also regulated *where* Israel did worship.

Post-conquest, God would choose a place in Canaan where He would especially dwell. While the entire land would be an Eden-like temple wherein God would live with His people, there was a particular location within Canaan where His theophanic presence would be manifest. This place, writes Gary Millar, would be "the land in microcosm.... the place within the place, where the full benefits of the covenant relationship were to be enjoyed."[7]

Moses didn't explicitly spell out the precise place, but the Israelites to whom he preached understood exactly what He was talking about. For forty years had passed since they left Sinai, and throughout their wilderness journey God had willed to meet with them at a particular place—the tabernacle. It was called "the tent of meeting" for a reason (Ex. 29:43). God had consistently met with Israel in the wilderness here, assembling them for public worship under the leadership of His priests (Lev. 8:3; 9:5; Num. 10:3). He had chosen to put His name upon and to reveal His glory at a portable tent, and when His people entered Canaan, that moveable dwelling would be set up in Shiloh (Josh. 18:1). It would eventually be replaced by the non-portable temple, constructed under the royal oversight of David's son Solomon upon Mount Zion in Jerusalem (1 Kgs. 6–8). This would be God's chosen dwelling place where He willed to meet with His gathered people. Thus, Israel needed to ensure they offered their sacrifices in the right place: "Take care that you do not offer your burnt offerings at any place that

---

7. J. Garry Millar, *Now Choose Life: Theology and Ethics in Deuteronomy*, New Studies in Biblical Theology 6, ed. D. A. Carson (Downers Grove, IL: InterVarsity, 1998), 102–3.

you see, but at the place that the LORD will choose in one of your tribes, there you shall offer your burnt offerings, and there you shall do all that I am commanding you" (Deut. 12:13–14).

Failure here would result in idolatry. If Israel offered the right sacrifices at the right times, but in the wrong place, it would be *will* worship, not *worthy* worship. Thus, throughout his preaching of the Second Commandment, Moses stressed the place of sacrifice (Deut. 12:5, 11, 13–14).

It was a necessary emphasis because God was about to enlarge Israel's territory (Deut. 12:20). As the nation spread out geographically, the trek to the tabernacle-temple would become more difficult and inconvenient for many within Israel. Furthermore, the idol-worshiping nations they were about to drive out of Canaan had altars all over the place. Israel would be tempted to use those localized shrines to worship God instead of making the long trek to Jerusalem (Deut. 12:2–4, 20–27). Put differently, they would be tempted to neglect God's chosen place of worship for a self-chosen place of worship, all out of convenience. "Why travel all the way to the temple, when we could offer the same sacrifices in our own backyard?"

Does that sound familiar? We can now offer up to God His prescribed spiritual sacrifices without going anywhere. Livestream technology has streamlined the church's worship, making it scarily effortless and convenient. Instead of seeking out and going to a physical place, all that is needed is to seek out and go to a virtual "place." From our living room couch, we can hear preaching and be led in singing and prayer. But in so doing, have we neglected God's chosen place of new covenant worship?

God's Chosen Place Under the New Covenant

It is true that God's worship is no longer confined to one geographical location. Such was Christ's unambiguous message to the Samaritan woman at Jacob's well: "Woman, believe me, the hour is coming when neither on this mountain nor in Jerusalem will you worship the Father" (Jn. 4:21).

After the reign of Solomon, the nation of Israel divided (1 Kgs. 12). When the northern kingdom suffered exile at the hands of Assyria in 722 B.C., pagans settled into the land and intermarried with the surviving Israelites. The result was an ethnically mixed people marked by religious syncretism who came to be known as the Samaritans. They had their own Scriptures (a revised version of the Pentateuch) and their own holy mountain (Mount Gerizim).

While the Jews and the Samaritans disagreed on most things, the one matter they agreed upon was that formal worship was to be carried out at a particular geographical location. They just couldn't agree on the location! So, when the Samaritan woman found her sexually immoral lifestyle exposed by this Jewish rabbi, she quickly diverted the conversation to this long-standing disagreement over the place of worship (Jn. 4:16–20). Jesus' response was not what she expected. He didn't give a reasoned defense of the Jewish position, proving that it was indeed God's will for His worship to be carried out on Mount Zion in Jerusalem. Such would have been a legitimate response. But Jesus wasn't interested in making this woman a Jew; He was interested in making her a genuine worshiper through His soul-satisfying, life-giving salvation. So He informed her of a major shift on the historical horizon when worship would no longer be relegated to a single geographical location: "But the hour is coming, and is now here, when the true worshipers will worship the Father in spirit and truth, for the Father is seeking such people to worship him" (Jn. 4:23).

Under the types and shadows of the old covenant, God willed for a physical structure to be the place His people assembled for His worship. But, from the garden of Eden onward, it had always been His will for the entire earth to be a temple filled with image-bearing priests who lived to exalt Him.[8] Adam had failed to spread God's worship, and so too had Israel. But through the life, death, resurrection, and ascension of Christ, a new age dawned wherein the truth of Scripture would be worked into the human spirit by the Holy Spirit so that a vast multitude of sinners might affectionately exalt God in every place.

New covenant saints still have a holy mountain to climb in God's worship, but it is not a mountain you will find in a world atlas. "But you have come to Mount Zion and to the city of the living God, the heavenly Jerusalem" (Heb. 12:22). Israel's earthly mount with its physical temple was a copy of the heavenly mount with its heavenly temple (Heb. 8:5; 9:23). Having trampled upon the serpent's head, the righteous Christ ascended the holy mount into the heavenly Jerusalem to take His seat upon the throne of David in the presence of His Father. Having purified us by His blood and empowered us by His Spirit, we now have the ability to ascend the heavenly mountain by faith, joining the angels and glorified saints in their ceaseless worship of the Triune God (Rev. 4–5). According to Jesus, we can do that from any place on earth. No pilgrimage is required.

---

8. See, for example, G. K. Beale, *The Temple and the Church's Mission: A Biblical Theology of the Dwelling Place of God*, New Studies in Biblical Theology 17, ed. D. A. Carson (Downers Grove, IL: InterVarsity, 2004), 81–167.

Wouldn't that then mean that God no longer has a chosen place on earth for His worship? Such is the conclusion often drawn, especially when arguing for the use of livestream technology in Christian worship. John 4 is haphazardly quoted to assert that God no longer cares about the specific place in which He is worshiped.

But that couldn't be further from the truth. When Christ's words to the Samaritan woman are understood in the broader context of redemptive history, it is clear that Jesus is not saying the place of worship no longer matters to God, but rather that "the place of worship is no longer geographical but ecclesial."[9] God still has a tabernacle-temple on earth, it's just not a temple made of physical stones and confined to a physical piece of territory. As the Apostle Paul tells us, "In [Christ] you also are being built together into a dwelling place for God by the Spirit" (Eph. 2:22). Or as the Apostle Peter puts it, "You yourselves like living stones are being built up as a spiritual house, to be a holy priesthood, to offer spiritual sacrifices acceptable to God through Jesus Christ" (1 Pet. 2:5). The "you" in both passages are plural, referring to the believes who together comprise the particular churches being written to (Eph. 1:1; 1 Pet. 1:1–2). To reference Murray's description of *ekklesia* again, it is "the called-together ones" who form God's dwelling place on earth under the new covenant. Yes, it is wonderfully true that individual believers, through their union with Christ, become the dwelling place of God (1 Cor. 6:19). But the emphasis of the

9.   J. Ligon Duncan III, "Does God Care How We Worship?" in *Give Praise to God: A Vision for Reforming Worship,* eds. Philip Graham Ryken, Derek W. H. Thomas, and J. Ligon Duncan III (Philipsburg, NJ: Presbyterian & Reformed, 2003), 43.

New Testament is not on the individual, but on the corporate.[10] No longer do God's holy people assemble *at* the temple, for the hour has now come when they assemble *as* the temple. What a wonderous mystery! "Particular visible churches under visible pastors," writes Richard Sibbes, "now are God's tabernacle."[11]

Does God care where we worship under the new covenant? You better believe He does! It is the gathered church filled with the Spirit and word of Christ that is His chosen place. The geographical location of that gathering is irrelevant, *but the gathering itself is essential.* While God wills for individual Christians to worship Him privately and Christian families to worship Him domestically (just as He did under the old covenant), His ultimate goal in redeeming us is to make us a worshiping assembly "filled with all the fullness of God" (Eph. 3:19). That is what it means to be the church.

---

10. Before Paul refers to individual Christians in Corinth as God's temple (1 Cor. 6:19), he first refers to the church at Corinth in the plural as God's singular temple: "Do you not know that you [plural] are God's temple [singular] and that God's Spirit dwells in you [plural]?" (1 Cor. 3:16). In fact, I would argue that when Paul speaks of individuals as the temple in ch. 6, he is actually saying, "You can't go the way of sexual immorality because you, in body and in soul, have been made *a part* of God's latter day temple!" He is stressing the individual as he warns against particular sins, but he is doing so under the controlling assumption of the corporate and communal.

11. Quoted in C. H. Spurgeon, *The Treasury of David*, vol. 1, *Psalms 1-26* (Grand Rapids: Zondervan, 1968), 10. Benjamin Gladd and Matthew Harmon explain, "Because God's latter-day presence has descended on his people through the outpouring of the Spirit, God's heavenly temple has extended down to earth and encompassed his eschatological people" [*Making All Things New: Inaugurated Eschatology for the Life of the Church* (Grand Rapids: Baker Academic, 2016), 130].

The Primacy of Public Worship

David Clarkson, the pastoral successor of John Owen, once preached a sermon on the opening words of Psalm 87: "On the holy mount stands the city he founded; the LORD loves the gates of Zion more than all the dwelling places of Jacob." Given that worship was to be carried out by individual families in Israel (Deut. 6:7; Josh. 24:15), what was it about the worship carried out at Zion that made it superiorly lovely in the eyes of God? Clarkson draws the conclusion that it was because "the worship of God in the gates of Zion was *public*," whereas "His worship in the dwellings of Jacob was *private*."[12] In typical Puritan fashion, he provides extensive biblical argumentation under twelve points as to why God values public worship more than private or family worship.[13] For our purposes I will summarize His points in my own words under three points, though I strongly commend to you the entire sermon.

First, public worship brings greater glory to God. One Christian may play the music of the redeemed on their own, but when the members of the church wield their distinct instruments (figuratively speaking) in harmonious unison, they are able to praise God for His greatness and grace in a way that no solitary Christian ever could. Furthermore, in the corporate assembly God's exaltation is made public, declaring to onlookers the infinite worth of God.

Second, public worship brings greater good to the church. It is in the public assembly of His people that God delights to dwell in His saving and sanctifying grace. Here is the primary

---

12. David Clarkson, *Prizing Public Worship*, ed. Jonathan Landry Cruse (Grand Rapids: Reformation Heritage Books, 2023), 4.

13. Clarkson, *Prizing Public Worship*, 11–34.

place Christ shepherds His sheep, guiding them to soul-nourishing pastures and guarding them from soul-destroying apostacy as they engage communally in word, sacrament, prayer, and praise. While one may be greatly edified reading and meditating upon the Scriptures in private worship, it is in the public reading and preaching of the word that Christ does His greatest works of spiritual transformation.

Finally, public worship brings us closest to God's goal for history. The limited glimpses we are given into the heavenly Jerusalem reveal that it does not consist of a multitude of perfected saints and angels who each glorify and enjoy God in their own personal monastic cells. The worship of heaven is pervasively public and communal. It is the worship of an assembly that defies the ability to count! This has always been God's goal for history, and we are given a foretaste of it in the public assembly of the local church. We are never closer to heaven than when we are gathered as God's called-out and called-together people to affectionately exalt Him through Christ and by the Spirit.

In our individualistic age, it is common for Christians to think of public worship as an optional extra. Private devotions are assumed to be the big thing. But without belittling private or family worship, biblical religion, under both administrations of the covenant of grace, teaches that public worship is the big thing. As Jonty Rhodes beautifully states, "Corporate worship is the sun around which personal and family devotions rotate."[14] The sacred assembly ought to have primacy in our

---

14. Jonty Rhodes, *Reformed Worship* (Phillipsburg, NJ: Presbyterian & Reformed, 2023), 75.

thinking, desiring, and living as God's called-together people. For, James Bannerman contends,

> All the parts of Church worship belong in a peculiar and emphatic sense to the Church, and they are made effectual by the presence and Spirit of Christ, as His instruments for building up and strengthening the collective body of believers in a manner and to an extent unknown in the case of private and solitary worship. The outward provision which Christ has made for social Christianity, as embodied and realized in the communion of the Church, is richer in grace and more abundant in blessing by far than the provision made for individual Christianity, as embodied and realized in separate believers. The positive institutions of Church worship, designed for Christians associated in a Church state, carry with them a virtue unknown in the case of Christians individually.[15]

Woe to the generation of Christians that forgets this!

But Is It a Valid Distinction?

Unfortunately, there is a lot of muddled thinking in our day about the distinction between public worship and private worship. For example, John Frame, arguing that all of life is to be worship, rejects the idea that "there is a sharp distinction between what we do in the meeting and what we do outside of it....All the earth is God's temple."[16] Certainly, God is in every place (Ps. 139:7–12), and we are to do everything unto His

---

15. James Bannerman, *The Church of Christ: A Treatise on the Nature, Powers, Ordinances, Discipline, and Government of the Christian Church* (Edinburgh: Banner of Truth, 2015), 347.

16. John M. Frame, *Worship in Spirit and Truth* (Phillipsburg, NJ: Presbyterian & Reformed, 1996), 34.

glory (1 Cor. 10:31). We can agree further that it is God's will to make this entire earth His holy of holies, full of perfected priests who live for His praise (Rev. 21–22). But the Scriptures make plain that God's temple, while presently expanding on the earth through gospel advance, is not synonymous with the earth. God's dwelling place, as we have seen, is the people bound together through their union with the exalted Christ. Those people are to live worshipful lives, but there is a sharp distinction between living unto God in the home and workplace and worshiping God in the public gathering of His people, and it is a distinction revealed not only in the Old Testament but in the New.

Admittedly, the New Testament authors do not speak much to the public worship of the local church. The book of Acts gives us windows into the corporate gatherings of the saints (Acts 2:42–47; 4:23–37; 6:1–7; 13:1–3; 14:21–23, 27; 20:7–12), but the thrust of these historical narratives is descriptive, not prescriptive.[17] It may be evident from these ac-

---

17. Furthermore, Luke only refers to the activity of the gathered church as "worship" once in his second volume (Acts 13:2). This was likely to distance the public worship of the church from idolatrous religious gatherings. David Peterson explains that Luke "restricted the term [*proskynein*] to a quite technical usage, applying it to those engaged on a pilgrimage to honour God in the traditional temple services (Acts 8:27; 24:11, *cf.* Jn. 12:20) or to the practice of idolatry (Acts 7:43, adapting Am. 5:26). Such terminology was presumably not applied to Christian meetings in Acts or the epistles because of its particular association with the rites of paganism or with the Jewish cult centred at Jerusalem" [*Engaging with God: A Biblical Theology of Worship* (Downers Grove, IL: IVP Academic, 1992), 148]. Rather than using the term *worship*, Luke focuses on the particular elements the early church engaged in as it gathered together to worship, distancing it from the false worship of unbelieving Judaism and paganism.

counts that the early church gathered together for worship and fellowship, but can it therefore be concluded that what took place in their public assemblies was understood to be distinct from what took place in their private lives and families? Some claim that the lack of explicit treatment on public worship in the New Testament is proof that, while the church did indeed gather, the public-private distinction, valid under the old covenant, has ceased to be a reality (or at least, has lost the sharpness of its differentiation). "Yet," as Edmund Clowney wisely retorts, "there is always the danger of disregarding things that are not said in Scripture simply because they did not need to be. They were the things that the readers took for granted."[18] Jewish converts would certainly have assumed the public-private distinction given the nature of old covenant religion, but so too would Gentile converts given the nature of pagan religion with its holy places and communal and corporate gatherings. For this reason the New Testament authors saw no need to write it large over their inspired testimony. It was all too obvious to everyone. This doesn't mean the New Testament doesn't explicitly distinguish between public and private worship. It does! But it does so in a way we would expect if it were indeed being taken for granted. The distinction arises as a grand assumption in need of no argumentation.

In 1 Corinthians 11–14, the apostle addresses three controversies that had developed in Corinth: public prayer and the covering of the head (11:2–16), the public partaking of the sacramental meal (11:17–34), and the public use of spiritual

---

18. Edmund P. Clowney, "Corporate Worship: A Means of Grace," in *Give Praise to God: A Vision for Reforming Worship,* eds. Philip Graham Ryken, Derek W. H. Thomas, and J. Ligon Duncan III (Philipsburg, NJ: Presbyterian & Reformed, 2003), 98.

gifts, specifically speaking gifts (12:1–14:40). As the apostle addresses these particular errors, the undergirding assumption throughout is that local churches gather regularly for worship via word, sacrament, and prayer. With regards to head coverings, he appeals to the practice of "the churches [i.e., the assemblies] of God" (11:16). Regarding the Lord's Supper, Paul stresses five times that it is partaken of, not individually, but "when you come together" (11:17, 18, 20, 33, 34). His words in v. 18 are particularly pertinent: "For, in the first place, when you come together as a church [i.e., an assembly]…" Geerhardus Vos rightly translates Paul's words as, "When you come together as the gathering of believers."[19] Furthermore, Paul explains that there are particular kinds of eating and drinking that are appropriate for believers to do in their private dwellings that are altogether inappropriate in the public gathering of the saints: "if anyone is hungry, let him eat at home—so that when you come together it will not be for judgment" (v. 34; cf. v. 22). Is the eating at home unto the glory of God? Absolutely. Paul just stated that it ought to be so (10:31). But the apostle took for granted the sharp distinction between worshipful living in private life and worship in the public assembly. He continues to do so as he addresses the use of spiritual gifts which are given by Christ "for the common good" (12:7) as each member uses his or her unique gifts (similar to the diverse instruments in an orchestra) to serve God and others (12:8–25). Without delving into whether or not these revelatory gifts continue in the post-apostolic era, it is clear that Paul's teaching on tongues and prophecy assumes a common assembly. Prophecy "builds

---

19. Geerhardus Vos, *Reformed Dogmatics*, vol. 5, *Ecclesiology, The Means of Grace, Eschatology*, trans. and ed. Richard B. Gaffin Jr. (Bellingham, WA: Lexham, 2016), 12.

up the church [i.e., the assembly]" (14:4), and tongues requires an interpretation "so that the church [i.e., the assembly] may be built up" (14:5). The Corinthians are to use their spiritual gifts to "strive to excel in building up the church [i.e., the assembly]" (14:12). Speaking in tongues might be personally edifying, but Paul confesses that "in church [i.e., in assembly] I would rather speak five words with my mind in order to instruct others, than ten thousand words in a tongue" (14:19). He refers again to when "the whole church [i.e. assembly] comes together" (14:23). Similar to his statements on the Lord's Supper, he makes clear that there are some forms of speaking that are appropriate in private worship or family worship that are not appropriate in public worship (14:27, 33–35).

The unique problems in the Corinthian church reveal that Christ's inspired spokesman takes the public-private distinction for granted. Herman Ridderbos puts it well: "However much the 'liturgy' must be seen as a spiritual worship of God embracing the whole of life (Rom. 12:1, 2), this does not alter the fact that the indwelling in and communion of Christ with the church have their point of concentration and special realization in its unity as assembled congregation."[20] Arguably, the public-private distinction isn't merely assumed, it is essential.[21]

Asking Some Difficult Questions

To summarize what we have seen up to this point, God wills for His worship to be carried out in a particular place, namely the public assembly of His called-together people. "As always from

---

20. Herman Ridderbos, *Paul: An Outline of His Theology*, trans. John Richard De Witt (Grand Rapids: Eerdmans, 1975), 486.

21. Terry L. Johnson, *Reformed Worship: Worship That Is According to Scripture* (England: Evangelical Press, 2015), 10–14.

the foundation of the world, so in the New Testament, the solemn worship of God is to be performed in the assemblies of his saints."[22] Assembling is the essence of being the church, and assembling is the essential precondition for the church to carry out her primary vocation of worship. These vital truths give rise to two difficult questions we must ask when considering whether livestream is a valid medium for the church's worship.

First, is it possible to assemble without assembling? When Paul, for example, speaks of the believers in Corinth *coming together* for worship, is it possible to do that without being physically with one another? Can we assemble virtually? Until quite recently, assembling always required physically going to a particular place. Asaph confessed that he was nearly consumed with envy "until [he] went into the sanctuary of God" (Ps. 73:17). David's heart was filled with gladness when his companions suggested, "Let us go to the house of the LORD!" (Ps. 122:1). A temple musician, lamenting his distance from Zion, recalled "how [he] would go with the throng and lead them in procession to the house of God with glad shouts and songs of praise, a multitude keeping festival" (Ps. 42:4). The psalms consistently call God's people to come into the temple courts for worship (Ps. 96:8; 134:1; 150:1). The obvious assumption in all of this is that the worshiper must leave his or her private dwelling in order to travel to God's dwelling "in the company of the upright, in the congregation" (Ps. 111:1). Under the old covenant, there was no assembling without physically gathering as an assembly. But under the new covenant, it is perhaps not so clear cut. For by the Spirit and faith,

---

22. John Owen, "The Nature and Beauty of Gospel Worship," in *The Works of John Owen*, ed. William H. Goold (Edinburgh: Banner of Truth, 1965), 9:74.

we ascend the heavenly mount to join the heavenly assembly, even while physically on earth (Heb. 12:22–24). There is an assembly that we enjoy now, though our embodied existence is not yet there. But God has willed for us to enter into that heavenly assembly as we gather in earthly assemblies, and the New Testament presupposes that those ecclesiastical gatherings are necessarily embodied, requiring physically coming together at a particular place. It's hard to imagine that God would consider physical distant believers connecting through a shared server and the mediation of individual screens to be a legitimate assembly in the biblical sense. The idea of a virtual assembly seems to be less of an oxymoron and more of an outright contradiction of terms.

We must ask, second, is it possible for something public to be experienced in private? When a church streams a worship service, it is broadcasted online to the general public. In that sense, anyone watching it could be said to experience what is public from the privacy of their homes. But when we talk about worship being public, we are not referring to something made available to the general public, but to something experienced and done in the public assembly of God's people. What is the fundamental difference between public and private worship? The one engages with God *in the church*, while the other engages with God *in the closet*. It goes without saying that you can't experience public worship from the privacy of your home.[23] I would suggest that when someone streams a worship service from the living room, they are not partaking in public worship. At best, they are using a public worship service to assist them in private worship (if they are streaming as an individual) or fam-

---

23. Unless, of course, your home is the place where the church as a corporate body of believers gathers!

ily worship (if they are streaming as a family). There is nothing wrong with being assisted in private and family worship. I will often read a devotional writing or even listen to a recorded sermon to stir up my heart in private worship, and sometimes I will use such material to assist me in leading family worship. The problem is assuming that streaming a service by myself is more than private worship. When we begin to believe we are engaging in public worship when there is nobody in the room but ourselves and God, we have embraced an illusion. It is an illusion that church leaders unintentionally promote when they propose that we can join them for worship while still in private. I realize those are strong words, but are they not true? What are we communicating when we promote this medium for public worship? How are we shaping people's perceptions of the nature and necessity of Christ's assembly?

The Apostle Paul shows us that there are some things that are legitimate in certain areas of life that are inappropriate and unbefitting of the gathered church's public worship. I'm thankful for the ability to Facetime with family members and to have Zoom meetings with friends who live in other parts of the world. I'm thankful to be able to listen to sermons throughout the week with the click of a button on my laptop. Praise God for these gifts mediated to us through recent technological innovation! But what is appropriate and fitting for me to do in my private life, may be entirely inappropriate and unbefitting in the assembly of the church for public worship.

Given the teaching of the Scriptures on the necessity of the corporate assembly, it is hard to see how the medium of livestream does not strike at the very substance of public worship. This form is not neutral and appears incompatible with the scriptural teaching on the necessity of the assembly unless, of course, we radically redefine what it means to assemble, which will be the focus of the next chapter.

Hindered and Homebound

At this point, it should be clear that a Christian cannot be a homebody, willfully choosing to stay home when he or she is physically able to assemble with the church. To replace the public gathering for a livestream service when you have the ability to gather is in direct contradiction to the will of God. To favor virtual "church" over the embodied assembly is a serious sin. If you are a Christian homebody, the call is simple—repent and return.

But what about the homebound, who favor the physical gathering of the saints but are physically unable to join in? By far the weightiest argument in favor of livestreaming church services is its apparent ability to connect and encourage church members who are providentially hindered from assembling due to poor health. In fact, it is the only pro-livestream argument that exercises significant force upon my soul.[24] I am a pastor who loves my people profoundly, and it grieves me when they are homebound, especially for extended periods of time.

I remember distinctly the elders' meeting in which we decided to pull the plug on livestream in 2023. There was an elderly woman in the congregation who had been homebound for some time, and it didn't look like Jean would be physically able to return to worship anytime soon. The forcefulness of the biblical and theological arguments against livestream led the leaders of Cornerstone to stop using this technological medi-

---

24. We will look at other pro-livestream arguments in pages ahead, including the idea that livestream is a great tool to serve the unchurched and to evangelize the lost.

um in public worship (since theology must always drive pastoral practice), but not without lamenting what that meant for our sister Jean. We discussed how to break the news to her, and how to ensure we shepherded her well through her physical debilitation. But the news never reached her. For the morning after that meeting, I found myself holding her hand, tearfully reading Scripture and praying with her, as an unconscious Jean prepared to pass into glory. She died that day, joining "the spirits of the righteous made perfect" in the heavenly assembly (Heb. 12:23).

Great care is needed when reading providence, but I read Jean's death as God's way of saying, "I don't need your innovative intrusions into my worship to preserve my suffering saints and get them to glory. Your job, Nick, is to lead with biblically-informed convictions. Leave the rest to me."

There have always been saints who have been providentially hindered from gathering in God's chosen place, sometimes for years at a time. Historically, love toward shut-ins led the church to go to them, so that they could experience, at least in part, what the entire congregation experienced in gathering.

Elders, together with other members of the church, should regularly visit the homebound, especially on the Lord's Day and often with word and sacrament. This gives them more of a genuine taste of the assembling of the saints and the public ordinances, all without potentially dulling their spiritual senses to the essentiality of the assembly. It is far easier for the elders to give a shut-in a link to livestream, than to commit to physically visiting a shut-in on a regular basis to minister the word of God to their soul.[25]

---

25. Some might object that this is a false dichotomy and that the elders could both physically visit the homebound on the Lord's Day and pro-

Lamenting as an Alternative to Livestreaming

When our churches were not gathering in the beginning months of the Covid pandemic, the great debate, particularly among Reformed churches, was over the legitimacy of administering the Lord's Supper virtually. Many rightly stood against this (we will explore why in the next chapter), arguing that in this extraordinary circumstance in the life of the church we ought not to introduce innovations into God's worship that threaten to undermine it.

What did they propose in the place of virtual communion? Lamentation.

Harrison Perkins exhorted the church, "Rather than feeling normal about our present circumstances by pretending that we can receive God's ordinary means of grace over the internet, we should pray vigorously for God to end the present crisis."[26] Scott Swain wrote, "In situations of loss such as this one [the loss of the ability to assemble to partake of the sacrament], we must learn how to lament, and we must learn how to teach

---

vide them with a link to the livestreamed service. But in my experience, along with falsely communicating to the homebound that they are participating in public worship (when, in fact, they are not), the livestream link also gives the elders an excuse to not take pains to physically go to the shut-in ("Well, at least they are able to livestream the services!"). Every body of elders has to decide how best to shepherd the homebound under their care, and as I will argue in the pages ahead, some may choose to make livestream available for shut-ins as one way to care for them pastorally. But at the very least, a livestream link cannot be a replacement for the elders physically and frequently going to the homebound with word and sacrament on God's holy day.

26. Harrison Perkins, "Virtual Communion," *Modern Reformation,* June 3, 2020.

God's people to lament, something quite difficult for those (like me) who are accustomed to instant gratification."[27] These men rightly noted that it was better to forgo the sacramental element than to attempt to partake of it through a medium that was inimical to the element. The appropriate response in such a grievous time is to grieve and cry to God for deliverance. What has been confusing to me, however, is why we have been quick to make this argument about one of the public means of grace, but not all of them. For example, in an article written during the pandemic, Greg Reynolds comments concerning the Lord's Supper, "It is impossible, *along with the rest of worship*, to administer electronically" (emphasis mine).[28] Do we actually believe that? If so, are we not communicating a different message when we allow people to stream our services electronically? I would propose that the same prescribed response to the loss of communion by Perkins and Swain ought to be the prescribed response for public worship in general, lest (to use the words of Perkins) we feel "normal about our present circumstances by pretending that we can receive God's ordinary means of grace over the internet." I realize that may sound harsh. But if the theological convictions articulated up to this point are true, is there anything that could be more loving and pastoral?

I would submit to you that this is the way laid out for us in the Psalms. Take Psalms 42 and 43 as examples. These psalms, which were likely a single composition originally, comprise the lament of a temple servant who had been driven into the wilderness by his enemies far from the house of God. His physical

---

27. Scott Swain, "Should we live stream the Lord's Supper?" *Reformed Blogmatics,* March 30, 2020.

28. Gregory Reynolds, "Virtual Church Meetings in a Time of Plague," accessed at https://opc.org/feature.html?feature_id=467.

distance from the earthly Zion left His soul ravenously desirous to meet with God: "My soul thirsts for God, for the living God. When shall I come and appear before God?" (Ps. 42:2). He knew God was present with Him in his wilderness trials which is why he prayed (vv. 8–9). But he also understood that there is all the difference in the world between experiencing God in a privatized wilderness and experiencing God with the sacred throng on Zion. He panted after the latter with agonizing groans: "Send out your light and your truth; let them lead me; let them bring me to your holy hill and to your dwelling! Then I will go to the altar of God, to God my exceeding joy, and I will praise you with the lyre, O God, my God!" (Ps. 43:3–4). This man wanted to physically go to the place where God was pleased to manifest His glory under the old covenant, because there is no greater joy than communion with God in His house through the blood-splattered altar. For the same reason, another exile, banished from the holy city expressed, "Let my tongue stick to the roof of my mouth, if I do not remember you, if I do not set Jerusalem above my highest joy!" (Ps. 137:6). The memories of past experiences in the temple combined with the present inability to gather gave birth to an agonizing desire to assemble in God's joy-inducing house.[29]

That prolonged temple-oriented agony is extremely painful, but it is also extraordinarily beautiful. In our digital age, we can easily forget that physical distance from the people and things we love is a means by which our love for those people and things only increases. We are quick to alleviate the pain,

---

29. For a fuller treatment of this intense longing for God's house, see chapter four of my forthcoming book *Restless Devotion: An Urgent Call to Godward Discontentment* (Grand Rapids: Reformation Heritage Books, 2025).

but what if the pain is actually a part of God's good purpose? What if it is actually His instrument to expand our hearts with holy desires and orient our hearts heavenward?

If the psalmist had the ability to get on his iPad and stream the sacred service in the temple, I doubt we would have these agonizingly beautiful songs. The first time he streamed temple worship he would have thought to himself, "That didn't come close to actually being there!" But by the hundred-and-first time, one look at his soul would likely reveal he had been slowly lulled into a sleepy contentment with his separation from the house of God.[30]

When my sheep are unable to gather for public worship, I want them to experience the agonizing groans of the psalmist. This is what a healthy Christian does when they or one they

---

30. This is a very different way of thinking than that of Matt Peeples who argues that those who livestream worship actually come to hunger for it more. In his article "Why Our Church Will Keep Livestream," Peeples writes, "We live in a highly mobile culture in which regular church attendance has dropped from once a week to once a month. Vacations, youth sports, and other priorities seem to pull on people more in our generation than in our parents' or grandparents' generations. By engaging online when they're away, they can develop the rhythm and habit of hearing and engaging with the Word. This regular engagement with the Word preached creates an increased hunger for it" (*The Gospel Coalition,* May 27, 2021). Without getting into how a pastor could be okay with the members of his church consistently attending public worship only once a month because of sports and other priorities, I think Peeples is mistaken to say that making livestream available will change that by growing peoples' hunger for the word. If anything, it sends the message that it is okay to not come and okay to have higher priorities than public worship on the Lord's Day because you can just stream it live or at a later time.

love is far from God's assembly. By giving people the illusion of being in assembly, livestream threatens to dull the spiritual desire for the assembly. I know from personal experience because it happened to me during the months of the Covid-19 lockdowns, and I never want to experience it again, nor do I want to be the instrument of dulling my people's affections for the irreplaceable, flesh-and-blood gathering of the church. That is not just the case with those homebound for months and years at a time but also with the family who comes down with a stomach bug. I don't want my people sitting at home sick on Sunday to be duped into thinking they are not missing all that much because of an artificial ecclesiastical experience mediated through a screen. When they wake up on Sunday morning and realize the flu renders them unable to go to God's house, I want it to give rise to an internal groan.[31] You can call me unloving and unpastoral, but I'm willing to be labeled such for the sake of true love.

We need to beware of introducing anything into the life of the church that would make the public assembly less than essential. And we need to wake up to the shaping power that livestream has to subtly, over the course of months and years,

---

31. In saying this, I'm not advocating that the sick do nothing but sit around and bemoan their circumstances when homebound on the Lord's Day. We ought to do everything we can to keep Sunday set apart unto God when physically unable to assemble with the church. How? By engaging in the private means of grace, particularly through meditation upon the word and prayer. In my experience, this is admittedly very difficult, especially with young children who are sick. But God understands our weakness and bears with us in it. We should do our best to seek and serve Him from the privacy of our homes and do so with an internal ache that recognizes that private (or family) worship is no replacement for public worship.

mold our hearts into a smug complacency toward the church and her worship, instead of regarding it as more precious than our highest earthly joy. For the public gathering of God's people is the place He has chosen to dwell and the place He wills to be worshiped. We cannot forgo it without striking at the substance of public worship itself, nor without striking at our identity as the church. The congregation that fully embeds its religious life and experience in a digital platform like Facebook loses the right to be called a congregation. For how can you have a church without a sacred assembly?

While our churches might not be entirely confined to an online server, we do need to ask what we are saying when we give people the option of "experiencing" church via streaming. Is our heart for the homebound and the lost leading us to embrace a medium that communicates the inessentiality of assembling?

It is a message that coheres remarkably well with the message of expressive individualism, and it is possible we are being shaped according to this radical secular vision even when watching our church's worship service from our tablet.

CHAPTER 3

BEWARE OF AN IMMATERIAL ANTHROPOLOGY

M Y FAMILY RECENTLY WORSHIPED with a confessionally Reformed church while on vacation. Before the call to worship went forth, a lady stood up to give the announcements. Greeting those sitting in front of her, she then proceeded to look directly at the back wall and said, "If you are watching us on livestream, thank you for coming!"

The sentence was revealing, and yet I wonder if anyone among the hundreds of people in the room even thought twice about it. Putting to the side the fact that public worship is not something we passively watch (we are worshipers, not consumers), this woman assumed that people had come to worship who in fact hadn't come. Up until recently, coming meant traveling to or entering into a place. But according to this announcement, you can now *come* without physically traveling anywhere as your disembodied existence is conveniently ushered into a distant place through a phone or laptop.

Language is constantly changing, and the way that past generations used words may not be the way our generation does. But there are some words worth fighting for, especially words with weighty biblical significance. *Come* is one of those words.

- "*Come*, O children, listen to me; I will teach you the fear of the LORD" (Ps. 34:11).
- "When shall I *come* and appear before God?" (Ps. 42:2).
- "I will *come* into your house with burnt offerings" (Ps. 66:13).
- "Let us *come* into his presence with thanksgiving" (Ps. 95:2).
- "Ascribe to the LORD the glory due his name; bring an offering, and *come* into his courts!" (Ps. 96:8).
- "*Come*, bless the LORD, all you servants of the LORD, who stand by night in the house of the LORD!" (Ps. 134:1).

When the psalmist uses the word *come* (in its Hebrew parallel, of course), there is no question that it entails human physicality. In fact, prior to the 21st century, this verbal action has always been understood to involve physical movement from one place to another. But our perception of what it means to come has now radically altered to the point that a person can be said to come without physically moving at all, and no one bats an eyelash.

It is a grand demonstration of the profound power of technology to shape the way we perceive reality and speak about it.

Extensions of Humanity

One of Marshall McLuhan's most helpful insights is that "media are extensions of some human faculty—psychic or physical."[1] The technological mediums we use are intended to expand our abilities as humans, and in this way they have the power to recast our understanding of humanity. As McLuhan explained, "Media, by altering the environment, evoke in us unique ratios of sense perceptions. The extension of any one sense alters the way we think and act—the way we perceive the world. When these ratios change, men change."[2]

One is left to wonder how a thinker like McLuhan would have assessed livestream. What human faculty would he have said it was an extension of? If he were living in the present, he might argue that livestream is an extension of the feet because it attempts to bring us places where we used to only be able to go on foot. Or he might argue that livestream is an extension of bodily presence because it seeks to extend our ability to exist in places that we never could have before. I don't have half the mind of McLuhan, but in my estimation, livestream is an extension of both. It extends the foot, altering our perception of what it means for humans to travel. It extends the body, altering our perception of what it means for humans to be present. That technological extension is the reason for the novel and popular belief that we can now come without coming.

Strangely, the technology that seeks to extend the body's ability to travel and to be present actually ends up devaluing bodily movement and presence by making them appear un-

---

1. Marshall McLuhan and Quentin Fiore, *The Medium is the Message: An Inventory of Effects* (Berkeley, CA: Gringko, 1967), 26.

2. McLuhan and Fiore, *The Medium is the Message*, 41.

necessary. There is no need to get out of bed and go anywhere when you can "come" simply with a swipe of your phone (still a bodily movement, mind you, but one that requires much less exertion). There is no need to physically enter into a place and fill a space because you can now be present from afar. In other words, the subtle message of this technological medium is that physical travel and presence are not important.

It is a message that coheres perfectly with the dualistic disregard for the human body that undergirds expressive individualism and the sexual revolution. Nancy Pearcy, in her book *Love Thy Body,* demonstrates how the anthropology undergirding secularism consists of a "fragmented, fractured, dualistic view of the human being" in which "denigration of the body is the unspoken assumption."[3] A separation of the human body from the human person "that involves a crassly utilitarian view of the body…along with a subjective, arbitrary definition of the person" is a unifying principle in our culture's propagation of abortion, euthanasia, fornication, homosexuality, transgenderism, and the redefinition (and forfeiture) of marriage and family.[4]

Take abortion as an example. Advances in medical technology have proven beyond a shadow of a doubt that even as a one-cell zygote the baby in the womb is a distinct biological human with his or her own unique genetic makeup. So how does our society justify killing such a one? By assuring us that at every stage of development (whether a zygote, embryo, or fetus) the baby is not yet a moral person with inherent dignity

---

3. Nancy R. Pearcy, *Love Thy Body: Answering Hard Questions about Life and Sexuality* (Grand Rapids: Baker, 2018), 19, 20.

4. Pearcy, *Love Thy Body,* 63. Her entire book fleshes out the unifying anthropological principle in relation to these pressing ethical matters.

and rights. The baby in-utero is a human body, we are told, but because it lacks self-awareness and certain cognitive abilities it is not a human person and so may be dismembered and discarded without murder being committed. There is an inherent "denigration of the body" at play and a divorce of the body from personhood.

Modern technology may be the greatest tool used by the devil to normalize such a body-degrading separation.[5]

Digitized Disembodiment

If you think that is an extreme statement, consider Samuel James's definition of the internet: "the disembodied electronic environment that we enter through connected devices for the purpose of accessing information, relationships, and media that are not available to us in a physical format."[6] Applying this more narrowly to livestream, this internet technology ushers us into a disembodied virtual domain. We are so used to it that we don't even think about it.

What is happening when you join the business meeting on Zoom? You are engaging in an electronic environment that has no place for your physical body and presence, though it will simulate your body and presence via tiny pixels on a screen. James is certainly correct when he states, "The internet, which dominates our lives as the primary medium through which we encounter most of the world, is an entirely disembodied habitat. Consequently, the internet trains our consciences to think

---

5. Pearcy notes, "The technocratic mindset celebrates unlimited dominion over the body and its functions" (*Love Thy Body*, 253–54).

6. Samuel James, *Digital Liturgies: Rediscovering Christian Wisdom in an Online Age* (Wheaton, IL: Crossway, 2023), 12.

of ourselves and the world in disembodied ways."[7] These digital technologies shape us to view our bodies as unnecessary and as hindrances that can be overcome with the click of a button.

It is not so much the content streamed via the internet that promotes an anti-body agenda. You might watch a sermon on the inherent goodness of the human body proven in the bodily resurrection of Christ and the believer's hope of future bodily resurrection in Him. The content might be pro-body, but the form of the internet itself promotes what James calls "a worldview of disembodiment."[8] To use this technology in an uncritical fashion primes us, and our children, to sell our souls to the dualistic, body-degrading idolatry of expressive individualism.

In our digital age, we have to fight for the verb *come*, lest we forget what it means to be human. To put it plainly, we cannot come without coming. We must resist the technological claims that assure us otherwise.

Our Intuitive Sense of the Body's Importance

I graduated from seminary in the spring of 2020 right in the midst of the Covid-19 lockdowns. For years I had envisioned the joy of graduation—a joy that would be shared with my closest family and friends. But those loved ones were not able to attend. Given the government restrictions, only graduates, professors, and a select few individuals, needed to run the ceremony, were allowed to be present. Even the main speaker sent a video recording of his message. It was anticlimactic, to say the least. But the ceremony was livestreamed so that my wife, boys, parents, in-laws, and friends could watch it. You could argue that they were still sharing in the event via screens, but

---

7. James, *Digital Liturgies,* 29.

8. James, *Digital Liturgies,* 29.

every graduate sitting in that empty room during the ceremony lamented the inability of those they loved to be physically present for the momentous occasion.

I didn't hold it against my wife for not being there because she was prohibited from coming. But imagine that my graduation was open to the public, and Tessa informed me the morning of, "I've had a pretty busy week, so I'm just going to stay home and livestream the graduation tonight. Know that I will be cheering you on from the couch!" I would have been deeply hurt. Her unwillingness to physically come would have inherently communicated that the occasion was not all that important to her.

For all our disregard for the physical body, every human knows intuitively that bodily presence is significant. You don't stream a wedding, a funeral, or a graduation of someone close to you unless there is absolutely no possible way of getting to the event physically. Why? Because being present at such significant events is important, and we recognize that we are not fully present without our bodies.[9] Furthermore, we recognize

---

9. The apostle Paul does claim in relation to the Corinthian church, "For though absent in body, I am present in spirit" (1 Cor. 5:3; cf. Col. 2:5). This is typically understood as an affirmation of Paul's oneness with the physically distant church in Corinth through the mutual indwelling of the Spirit. It is similar to his statement, "But since we were torn away from you, brothers, for a short time, *in person not in heart*, we endeavored the more eagerly and with great desire to see you face to face" (1 Thess. 2:17, emphasis mine). Paul was united to these Christians in heart or spirit, though separated from them in his embodied personhood, and that oneness in the Spirit made him long for the physical separation to be no more. While the apostle refers to a kind of spiritual presence and oneness, it is mediated by the Spirit, not by a digital device, and it in no way erodes the irreplaceability of physical presence

that we can't genuinely participate in such events without our bodies. Andy Crouch writes, "Only by showing up in person can we feel and grasp the full weight, joy, and vulnerability of the most important experiences in human life."[10]

If you value a particular event, you will go to great lengths to ensure you are able to be there physically, not merely through a disembodied screen. That is why the average price for a ticket to the 2024 Superbowl sold for over $8,000! People were willing to take out a second mortgage to be physically present for four hours at a football game, only to go on to pay $100 to park and $14 for a sub-par hotdog. Why? Because they recognized that there was a vast difference between watching the game on a screen and actually sitting in the stadium, and the game was important enough to them that they were willing to pay whatever was required to be genuinely present in bodily form.

When we have the ability to physically gather with Christ's people but choose to livestream instead, we are communicating where our values lie. This technological tool communicates to us the unimportance of the body, and our use of it communicates the unimportance of the event we are content to stream (in cases when we could actually be present).[11]

---

and proximity but instead fuels a longing after such. For a helpful treatment of these passages see Matthew Lee Anderson, *Earthen Vessels: Why Our Bodies Matter to Our Faith* (Grand Rapids: Bethany House, 2011), 212–18.

10. Andy Crouch, *The Tech-Wise Family: Everyday Steps for Putting Technology in Its Proper Place* (Grand Rapids: Baker, 2017), 199.

11. The argument could be made that those who are physically unable to attend an important event show that they value it by streaming it digitally (like Tessa watching my graduation ceremony from home). I agree

The healthy Christian recognizes that what is taking place in public worship is far more significant than a graduation ceremony. Assembling with the saints to ascend God's holy hill in worship is the most important thing we will ever do, and our willingness to be inconvenienced to partake of it physically is one way we convey our conviction of that. At God's command, we come in the only way that humans can come—with our bodies![12]

The Bodily Nature of Worship

God gave you a body in order that you might worship Him with it. In fact, the primary verbs translated *worship* in both Hebrew and Greek are inescapably physical. Daniel Block explains, "Both literally refer to subjects prostrated before a superior, a posture that states the equivalent of 'Long live the king.'"[13] This is why I define worship as affectionate, *whole-*

---

that in instances like a graduation ceremony, livestream can be a valuable tool to communicate we value a person or event. But as I argued in the previous chapter, public worship is in an entirely different category than any other human event, and it is hard to argue how streaming it doesn't undermine its necessarily public nature in a substantial way.

12. The only exception to this is the intermediate state as the spirits of the deceased come before God's heavenly throne, but the Scriptures present this disembodied condition as temporary and unnatural (2 Cor. 5:1–10), resulting in a longing for future resurrection glory and final judgment (Rev. 6:9–10).

13. Daniel I. Block, *For the Glory of God: Recovering a Biblical Theology of Worship* (Grand Rapids: Baker Academic, 2014), 12. Block persuasively refutes the idea that such external displays of devotion are no longer relevant for new covenant worship since it is spiritual, concluding, "Surely worship that pleases God involves bodily gestures of subordination and submission" (*For the Glory of God*, 15–17).

person exaltation. For worship is our communicative response to God's gracious communication to us, and we communicate what is in our hearts with our bodies. Such is true of both verbal and non-verbal communication. I cannot verbalize a syllable without using my lungs, vocal chords, teeth, tongue, and lips. While psychologists debate how much of our communication is non-verbal, it is clear that much of what we say comes from facial expressions, eye contact, physical posture, and bodily movements, and such is true of our communication toward God in His worship. Our bodies are not an optional extra, but a vital component of worship in private, family, and public.

Given the Creator-creature distinction, worship is best communicated via a physical posture that communicates His greatness and our consecrated allegiance to Him. The psalmist highlights this by way of synonymous parallelism in his call to worship: "Oh come, let us worship and bow down; let us kneel before the LORD, our Maker" (Ps. 95:6). We not only communicate our submission to God with our bodies in worship, but also our happy gladness in Him: "Clap your hands, all peoples! Shout to God with loud songs of joy!" (Ps. 47:1).

It was probably fifteen years ago when I first read C. S. Lewis's *Screwtape Letters,* a series of fictional letters from a senior demon to a demon in training. The book was riveting (and continues to be to this day), but I remember scratching my head when Lewis, through the lips of the senior demon Wormwood, stressed the importance of bodily posture in prayer:

> One of their poets, Coleridge, has recorded that he did not pray 'with moving lips and bended knees' but merely 'composed his spirit to love' and indulged 'a sense of supplication.' That is exactly the sort of prayer we want....At the very least, they can

be persuaded that the bodily position makes no difference to their prayers; for they constantly forget, what you must always remember, that they are animals and that whatever their bodies do affects their souls.[14]

In other places, Lewis makes clear that the body is not everything in prayer. "Kneeling does matter, but other things matter even more. A concentrated mind and a sitting body make for better prayer than a kneeling body and a mind half asleep."[15] The body is not everything, but the body is certainly something, and it is something God designed us to use as we affectionately exalt Him. Furthermore, as Lewis notes, what we do with our bodies affects our souls.

We are "a psychosomatic unity."[16] What we do with our psyche (i.e., our soul) will affect our soma (i.e., our body), and visa-versa. The body is an integral part of who we are, and it is an integral part of how we worship. As Gregg Allison writes, "Embodied worshipers properly render worship to God through whole-body devotion to him."[17]

The Second Commandment assumes that our worship will be externalized in concrete acts of devotion. Thomas Boston

---

14. C. S. Lewis, *The Screwtape Letters* (New York: Macmillan, 1976), 33–34.

15. C. S. Lewis, *Letters to Malcolm: Chiefly on Prayer* (San Diego: Harcourt, 1992), 17. For this reason, I typically go for a jog or pace back-and-forth during my morning prayer times as it helps me remain attentive. But I do so in the recognition that it is not the ideal bodily posture and that posture matters.

16. Anthony A. Hoekema, *Created in God's Image* (Grand Rapids: Eerdmans, 1986), 217.

17. Gregg R. Allison, *Embodied: Living as Whole People in a Fractured World* (Grand Rapids: Baker, 2021), 183–84.

makes this point, providing four arguments that "this com-mand...requires outward bodily worship":

1. "Is not God the God of the whole man, the body as well as the soul? Is it not highly reasonable, then, that we worship God with outward bodily worship, as well as with the inward worship of the heart?"

2. "God will not only be worshiped by us, but glorified before men....But our inward worship cannot do that, for that is what none can know but God and our own souls. Therefore outward worship is necessary."

3. "Out of the abundance of the heart the mouth spea-keth in other cases, and why not in this?...And though outward worship may be performed where there is no inward heart, yet if the heart be a temple to God, the smoke will rise up from the altar, and appear without in outward worship."

4. "Outward worship is not only a sign of the inward, but it is a help and furtherance to it."[18]

To engage with God in His worship is a fundamentally embodied reality for the creature. Yes, as new covenant priests we "offer spiritual sacrifices" to God (1 Pet. 2:5). But that does not mean non-physical sacrifices; it means sacrifices offered up in dependence upon the Holy Spirit. For who can offer up sacrifices of preaching, praying, praising, or partaking without flesh-and-blood physicality? Worship requires more than the

---

18. Thomas Boston, *An Illustration of the Doctrines of the Christian Religion*, in *The Complete Works of Thomas Boston* (Stoke-on-Trent, UK: Tent-maker, 2002), 2:140–41.

body, but not less than the body.[19] What is true of our coming, is equally true of what we do when we come.

We've seen that the form of livestream threatens the substance of public worship by gutting the adjective *public* of any substantial meaning. But what about the elements? Remember Duncan's warning that "the *forms* in which [the] elements are performed must not be inimical to the nature or content of the element."[20] Is livestreaming "inimical" to the non-negatable public means of grace?

Physical Communication and Connection in Preaching

Preaching, the God-exalting exposition and application of Scripture, is intensely physical. In my experience, there is nothing more simultaneously exhilarating and exhausting than proclaiming "the unsearchable riches of Christ" (Eph. 3:8). The exhilaration leads to the entire physical body of the preacher being utilized, and the entire physical body being utilized leads to exhaustion.

After a worship service one Sunday, I bent down to tie my shoe. As I stood up, a college student approached me, shook my hand, and said, "You're the only preacher I've ever seen who

---

19. See David Clarkson's soul-stirring exhortation at this point in *Prizing Public Worship*, ed. Jonathan Landry Cruse (Grand Rapids, MI: Reformation Heritage Books, 2023), 68. The central problem with old covenant Israel was that they drew near to God physically while their hearts were far from Him (Isa. 29:13). We certainly need to beware of this idolatrous error, but arguably a greater temptation for us in the 21st century West is to think our physicality is a matter of indifference so long as our hearts are engaged.

20. J Ligon Duncan III, "Foundations for Biblically Directed Worship," in *Give Praise to God*, 55–56.

moves so much in the pulpit that his shoes come untied!" I took that as a complement. For it is my conviction that it takes the whole preacher to proclaim the whole Christ—feet and all!

I don't practice my gestures and movements before preaching, nor am I thinking about them while I'm preaching. They are the spontaneous outworking of the truth of Christ dwelling richly in my heart. For when the law and gospel erupt from the depths of the soul, it won't merely come out in spoken language but also in what Robert Dabney calls "sign-language." My wife is deaf so I'm used to using facial expressions and hands to communicate. It is language that is seen, not heard. According to Dabney, such visual communication is a vital component of preaching:

> He who is master of this sign-language has, indeed, an almost magic power. When the orator can combine it with the spoken language, he acquires thereby exceeding vivacity of expression. Not only his mouth, but his eyes, his features, his fingers, speak. The hearers read the coming sentiment upon his countenance and limbs almost before his voice reaches their ears: they are both spectators and listeners; every sense is absorbed in charmed attention.[21]

Yes, even the fingers speak!

Every preacher is uniquely designed by God and will use his body according to his particular personality and physical constitution. But use his body he must. Our facial expressions, gestures, and other movements are a significant part of the communication taking place. John Piper recounts how an elderly woman in his congregation told him, "Pastor John, I just

---

21. Robert L. Dabney, *Sacred Rhetoric* (Edinburgh: Banner of Truth, 1979), 323.

love to watch you preach. I understand you because of your hands." He goes on to explain, "My left hand is moving as I'm trying to clarify what I'm saying, and it is just part of the energy, right? It is part of the expressiveness. It is the soul becoming visible and not just hearable."[22] Preaching is intended to be seen, not merely heard.

But doesn't livestream give us the ability to see the preacher? In a way, it does. Multi-site churches often broadcast the preacher on a screen during worship services at their various sites. They, however, always have a real flesh-and-blood person to lead the singing. Why? Because they recognize that watching someone lead congregational singing on a screen is not the same as being led in person.[23] If that is true of singing, how much more so is it true of preaching! It is not enough to have physical gestures and facial expressions replicated via digital pixels. For the physical presence itself is vital to receiving the full force of both verbal and non-verbal communication in preaching.

It is better experienced than explained, but there is a physical connection between the preacher and his hearers that is lost when the preacher is mediated through a digital medium. Part of that is because the congregation is also speaking to the preacher, sometimes verbally (e.g., "Amen!"), but always non-verbally with their eye contact, bodily posture, and facial expressions. The preacher needs that communication to do his work effectively. As Joshua McIlvaine states, "The mind of the speaker, e.g., being directed to his audience, his eye naturally

---

22. John Piper, "What's with All the Preaching Gestures?" *Ask Pastor John,* March 18, 2013.

23. This point has been made by many. See, for example, Christopher Ash, "Why I Object to Screen Preaching," *The Gospel Coalition,* April 16, 2013.

follows his mind—he looks at them; and not barely as 'a sea of faces,' without distinction, but he scans their individual countenances, notes their several expressions, and thus becomes conscious of the effect which he is producing upon them."[24] Moreover, when a pastor looks out at a people he knows and loves, it ignites a holy affection toward them and a desire to do them spiritual good. He sees the grieving widow and longs for her to know Christ's comfort. He sees the backslidden teenager and longs for him to be awakened to Christ's holy glory. He sees the children and longs to make gospel truth understandable so that they might embrace Christ. That cannot be replicated when preaching to a camera in an empty room.

Consider further that biblical preaching consists in reproving, rebuking, and exhorting (2 Tim. 4:2). It is generally unwise for reproof and correction to be given via text message or email. People need to hear the tone of your voice, but those mediums of communication don't allow for that (and thus shape how the message is received). Sometimes a phone call is the only option, but ideally you want to be face-to-face when issuing a rebuke. For people are more apt to receive hard words when they see your sympathetic facial expressions, gentle hand gestures, and the tears streaming down your cheeks. Again, that non-verbal communication can be virtually conveyed to a degree, but nothing can replace experiencing it in real time *and space*.

Can preaching be mediated through a screen? Sort of, but not really. For God has designed it to entail physical communication and connection which is significantly lost through the mode of livestream. At the very least, we can say that watching a sermon on a screen falls far short of the biblical ideal and de-

---

24. Joshua H. McIlvaine, *Elocution: The Source and Elements of Its Power* (New York: Charles Scribner's Sons, 1895), 98.

fies what common sense and universal human experience tells us about the nature of oratory.

Perceptive readers will likely have noticed a potential contradiction in my thinking and practice. For I have noted at various points that I am not opposed to the use of recorded sermons for personal edification. Visit our church website, and you will find links to past sermons preached from Cornerstone's pulpit. I wrestle with this, especially when I consider men like John Calvin, who were strongly opposed to having their sermons put into print because of their understanding of what a sermon was. But I also see the Holy Spirit moving the prophets and apostles to record sermons in the Scriptures for our edification and instruction (e.g., Jer. 7:1–15; Matt. 5–7; Acts 17:22–31). Reading those sermon records does not give us the illusion that we are participating in public worship, nor does it put itself forth as a potential replacement for public worship and the preaching that is at the center of it. I view sermon recordings made available online in a similar fashion. *In my mind there is a vast difference between telling people they can experience public worship in private via livestream and telling people they can be edified by a sermon or a song from public worship in private via a recording.* That being said, there is always the danger of a person's heart being molded to think they no longer need the local church and her worship. "I can find far more gifted preachers and musicians online than in any church near me. So why join a church?" There is also the danger of the preacher directing his preaching to some unknown digital audience rather than the known people entrusted to his spiritual care by Christ. These dangers give me pause, but I also see the benefits of making my sermons available to my people after the fact for their growth in grace.[25] So I'm hesitantly happy to live with the tension, not seeing it as a genuine contradiction.

The Sacraments Require Physical Presence

While the content of preaching may to some degree be mediated through a screen, such is not the case with the sacraments. Don't believe me? Try baptizing a physically distant person with water through a livestream connection on your iPad. It matters not what mode you prefer (sprinkling, pouring, or immersing), you will not be successful in getting a drop of water on the person, and you might just destroy your iPad!

Given that virtual baptism is obviously impossible (unless we affirm that people can baptize themselves or that "digital water" is a permissible baptismal form), we will give our attention to the Lord's Supper. During the Covid-19 pandemic many churches embraced the practice of "virtual communion." Each individual or family were responsible to get their own juice and bread in their homes, and the pastor would lead them in partaking via a screen.

Chris Ridgeway defends this practice in an article in *Christianity Today*, arguing that "daily digital culture has shaped our interactions to the point that *human presence is not synonymous to physicality.*" Instead of fighting for our embodied physical-

---

25. For example, a couple weeks ago I was on a home visit in which a member of the church informed me that she often relistens to my Sunday sermons the following week, journaling extensively about how the sermon applies to her personal life. She is present on Sunday to hear the word, and then listens to it again the next week with the intention of getting the word deeper. What a wonderful use of digital technology that doesn't undermine the public assembly and the public means of grace, but instead serves them!

ity as God's good design, Ridgeway resigns to what he calls a "technological presence." He explains,

> New technologies that first appear as toys (we play with them) soon turn into tools (we use them) and then become our technological territory—that assumed background environment wherein something like 'texting' becomes the conversation (or argument!). These 'environmental' technologies shift the focus from the tech back to the substance of human presence. Being present doesn't require being in person.[26]

He then goes on to apply this to the Supper, arguing that the cultural shift in how we understand human presence means we can now be communally present to partake of what is an essentially communal meal all without being physically present. He even goes so far as to liken our presence mediated through Zoom to the spiritual presence of Christ mediated through the bread and wine.[27] Here is further proof of the subtle power of digital technology to shape how we think about humanity, community, and even Christ.

Just because our culture redefines presence doesn't mean the church is free to do so. Marriage is now viewed as a contractual agreement between two consenting parties, but that does not warrant Christians altering their conviction that marriage is an exclusive covenant union between one man and one woman until death. So too, in our anti-body age, human presence may be redefined to no longer entail physical proximity, but we do not have the freedom to embrace that new idea if it contradicts God's thoughts about human presence. Does God care if we

---

26. Chris Ridgeway, "Online Communion Can Still Be Sacramental," *Christianity Today,* March 27, 2020.

27. Ridgeway, "Online Communion Can Still Be Sacramental."

are physically present in His physical assembly when we partake of the physical elements? Does He leave room for what we could call a "technological presence" when engaging with the sacraments? These are questions Ridgeway appears never to ask and certainly never answers.

When Jesus instituted the Supper, He did so in physical proximity to His disciples. He physically sat with them, physically pointed to the bread and cup saying, "*This* is my body and blood," physically gave the bread and cup to them, and told them He would not physically eat this meal with them again until the consummate marriage supper when He would feast with them in glorified human flesh (Luke 22:14–23). Instituting the Supper with the apostles, He entrusted it to them (and elders after them) as an abiding ordinance to be carried out in the church's worship (1 Cor. 11:23–25). It is to be partaken only when the saints come together in sacred assembly (1 Cor. 11:17, 18, 20, 33, 34). For how can the elders give the bread and cup to communicant members of the church without being physically present with them? They can't. Just as the waters of baptism cannot be applied through a screen, neither can the bread and wine be set apart by God's appointed leaders and distributed to the people through a livestream connection.[28]

Furthermore, the sacramental meal is a sign and seal of our communion, not only with Christ, but with one another. "Because there is one bread, we who are many are one body, for we

---

28. Returning to our discussion in chapter 1 on the keys of the kingdom, the elders have been given authority by Christ to welcome the repentant and believing to the Table, while also withholding the bread and wine from the unrepentant and unbelieving. But virtual communion makes this impossible, granting authority to each person watching to give themselves the bread and wine if they so please.

all partake of the one bread" (1 Cor. 10:17). United to Christ, we are united to each other, and we celebrate that in a common meal. Hence, there is a need to "discern the body," that is, to partake of the elements in the recognition of the gathered church and the preciousness of that gathering (1 Cor. 11:29). To settle for what Ridgeway labels "technological presence" is to lose the sacrament itself as a physical sign of our physical togetherness in Christ.

What about his argument pertaining to Christ's presence? It is true that Christ is really present, though His flesh-and-blood physicality is in heaven. But He is not present in an artificial way like pixels on a screen. Given His singularity of essence with the Holy Spirit and the way that through His ascension He has become "a life-giving spirit" (1 Cor. 15:45), He really is present as the Spirit of Christ fills the physical assembly and accompanies the word and sacrament. Calvin, denying the ubiquity of Christ's humanity, confessed that it is unlawful "to drag [Christ] from heaven" and attach him with, or in, the elements.[29] Yet, he said, the Spirit "surmounts all natural obstacles," enabling the believer to partake of the body and blood of Christ.[30] He does this, not by bringing the incarnate

29. John Calvin, *The Institutes of the Christian Religion*, ed. John T. McNeill, trans. Ford Lewis Battles, 2 vols., Library of Christian Classics 20-21 (Philadelphia: Westminster, 1960), 4.17.31. The *ubiquity* of Christ's humanity refers to the teaching that Christ's physical body can be in many places at once. While Luther and Lutheranism affirm this teaching, the Reformed have always denied it, arguing that the incarnate humanity of Christ is limited to a singular place, namely the right hand of the Father in the heavenly places.

30. John Calvin, "Clear Explanation of Sound Doctrine Concerning the True Partaking of the Flesh and Blood of Christ in the Holy Supper," in *Treatises on the Sacraments: Tracts by John Calvin*, trans. Henry Beve-

Christ down to us, but by lifting us up to Him.[31] The Spirit of Christ descends upon us in the sacrament to cause us to ascend to heavenly places to enjoy communion with the exalted and glorified God-man.[32]

This is a great mystery, and there will never be a technological correlation to it! Even if there was, God has made clear that the physical assembly is the only context in which the physical meal is to be received from the hands of physical elders and partaken with the physical mouths of Christ's physical people in physical proximity to one another. The form of livestream is entirely inimical to the sacramental elements.

What About the Other Elements?

In the divine-human dialogue of worship, God speaks to us by word and sacrament, and we respond in praise and prayer. Do we really need the physical assembly to praise and pray? It all depends. God wills private praise and prayer, but He also wills public praise and prayer. The one may be had in solitude, but the other cannot.

As it is in preaching and the sacraments, so too in our singing there is a horizonal dimension. Yes, we are singing vertically to God. "Oh sing to the LORD a new song; sing to the LORD, all the earth!" (Ps. 96:1). But we are also singing horizontally to one another. When the Spirit fills the assembly, it results in us "addressing one another in psalms and hymns and spiritual songs, singing and making melody to the Lord with [our hearts]" (Eph. 5:19). In the parallel passage in Colossians

ridge (Ross-shire: Christian Focus, 2002), 519.

31. Calvin, *Institutes*, 4.17.31.

32. Ronald S. Wallace, *Calvin's Doctrine of the Word and Sacrament* (Edinburgh: Oliver and Boyd, 1953), 206.

3:16, Paul explains that congregational singing in response to the word of Christ is a way of "teaching and admonishing one another in all wisdom." When we lift up our voices in song to extol God for His greatness and His grace, we are declaring that greatness and grace to each other. Through our singing, we are admonishing one another to cast away all idolatrous counterfeits and to taste and see the goodness of God.

You don't need impressive musicians to do that (in fact, impressive musicians often detract from that). All that is needed is a congregation that is gripped with the glory of God in Christ and sings of His glory with all of their hearts. The church I pastor has some very gifted musicians, but we keep the music simple so as not to detract from the singing itself. We often have visitors worshiping with us from out of town, and the number one comment I receive about our worship is, "Man, your church sings!" While my pride loves to have a visitor come up to me and say, "Man, you can preach!" the comments about our robust singing bring me great joy. They are evidence that the word of Christ, through the preaching, is dwelling in us richly by the Spirit's grace so that we cannot help but burst forth in joyous song. Through the horizontal exhortation of congregational singing God invades and overtakes our hearts with His weighty glory. Such cannot be mediated through a screen from your private dwelling. It requires flesh-and-blood assembling.

Out of all the public means of grace, prayer is perhaps the best fit for livestream. After all, we normally have our eyes closed and our heads bowed when we pray in public, only hearing the words and not seeing anyone or anything. I pray with people regularly over the phone or Zoom. I'm thankful for this. But it is still no replacement for being in the same room with brothers and sisters as we devote ourselves to "the prayers" (Acts 2:42). God wills His temple to be "a house of prayer for all the peoples" (Is. 56:7). He assured Solomon upon the completion of His old covenant house, "Now my eyes will be open and my ears attentive to the prayer that is made in this place" (2 Chron. 7:15). Surely, God heard the prayers offered up by individual Israelites in their private dwellings. But His ears were especially open to the prayers offered up in the corporate gatherings at His temple.

We are God's house as we unite together with our physical bodies to pour our hearts out to God in petitions that encompass not only ourselves but the nations (Ps. 67; Acts 4:23–31). It is when we are gathered together to pray that Christ particularly promises to hear and answer: "Again I say to you, if two of you agree on earth about anything they ask, it will be done for them by my Father in heaven. For where two or three are gathered in my name, there am I among them" (Matt. 18:19–20). I'd rather pray with a brother or sister over Zoom than not pray with them at all. But nothing can replace seeking the face of God via embodied gathering.

From the beginning of the worship service to the end, we need our bodies, and we need one another. In my tradition, we end the service with a benediction (upon the basis of the old covenant liturgy in Leviticus 9:22–24 and the new covenant epistles which often end with such). It is a very significant moment in the service as I stretch wide my hands and look my people directly in the eyes with a beaming countenance to pronounce God's blessing upon them. The benediction is a commission. By it, God freshly puts His triune name upon His gathered people and sends them forth into the world to be His representatives (Num. 6:22–27; 2 Cor. 13:14). You can't be sent out from the assembly back to your homes and work-places if you never left your home in the first place. Nor can the physical gestures and connection inherent to this blessing be adequately mediated through a screen.

Are you seeing how the elements of public worship are in-herently body-affirming and entail our physical presence to-gether? "Christianity," writes Samuel James, "does not reduce the self to the screen-mediated mind."[33] Thus, it does not, and cannot, reduce worshipers to a screen-mediated mind.

The form of livestream is at best a less-than-adequate form to carry out the public means of grace. This should give Chris-tians great pause in using it in the church's worship.

Embodied Encouragement

Let's return to the church service at the beginning of the chap-ter. Do you remember the greeting to those who were watching online? After the pastor gives the benediction and the service concludes, what do they do? They exit their web browser and go about their day. No fellowship with other believers. No op-portunity to be invited over for lunch. No avenue to know or be known.

Sherry Turkle has written a book about the effects of digital technology on our relationships which she fittingly titled *Alone Together.*[34] We've all seen it—the couple sitting together at the restaurant who are fixated on their phones as if the other is not there at all. Though physically together, technology leads them to act as if they are alone. Livestream flips that reality on its head. Instead of being alone together, when we stream a

---

33. James, *Digital Liturgies,* 110.

34. Sherry Turkle, *Alone Together: Why We Expect More from Technology and Less from Each Other* (New York: Basic Books, 2011).

church service we are *together alone*. Though physically alone, the screen simulates an experience that makes us feel together. But in reality, we are not.

It may surprise readers that we have gotten this far without quoting Hebrews 10:24–25, a go-to text for those who have concerns about livestreaming church: "And let us consider how to stir up one another to love and good works, *not neglecting to meet together*, as is the habit of some, but encouraging one another, and all the more as you see the Day drawing near" (emphasis mine). We are commanded to stir up and encourage one another in faith, hope, and love, and the logic of the writer to the Hebrews is that we cannot do that if we are "neglecting to meet together." Assembling is essential to the church's fellowship, even in our day of technological innovation.[35]

---

35. It could be argued that if a church is hosting virtual fellowship times through a shared server, then they are not neglecting to meet together. But the noun translated in Hebrews 10:25 as "meet together" is only used one other place in the New Testament. In 2 Thessalonians 2:1, Paul refers to the second coming of Christ "and our being *gathered together* to him." Just as Christ will not appear via hologram at the second coming but will physically descend from heaven in a glorified human body, so too all the saints will be supernaturally assembled before Him in physical, embodied proximity. No one will livestream the second coming! So too the encouragement of the church is designed to take place in an embodied gathering where saints bound together in Christ are in one another's physical presence. To argue for some kind of technological gathering in Hebrews 10:25 is to read into the text what is not there. To put it bluntly, virtually fellowship does not qualify as meeting together as a church. I'm indebted to my friend Mike Myers for the insightful connection between Hebrews 10:25 and 2 Thessalonians 2:1.

By our very physical presence in the assembly, we encourage one another. There are young parents who endure the difficulties of teaching their toddlers to sit in public worship. There are college students who slept in until 1:00pm on Saturday, but are present, attentive, and engaged in God's worship early on Sunday morning. There are the elderly who go to great lengths just to make it to their seat without being exhausted. When each of these members come despite all the inconveniences and discomforts of coming, they are saying to everyone else present, "Christ is worth it! Christ is risen! Christ is King!" They say it simply by showing up.

Then there is the physical conversation and embracing. One of my favorite parts of the Lord's Day is after the evening service when people linger, laughing together, crying together, confessing sin together, praying together, and simply enjoying being together. If you try to do that through a virtual habitat for interactive digital fellowship, it will flop. Brett McCracken writes, "In a lonely, disembodied world, the church offers a beautiful alternative: an enfleshed community where the manipulative filters of life online fall away and you can be known in a truer sense, warts and all."[36] There simply is no replacement for being physically together. God knows that, which is why He embedded it in His word for all time: "Don't neglect meeting together!"

On Cars and Church Life

A few weeks ago, we had a package accidentally delivered to our house that belonged to a neighbor about a quarter of a mile down the road. I had never met this neighbor, so when

---

36. Brett McCracken, *The Wisdom Pyramid: Feeding Your Soul in a Post-Truth World* (Wheaton, IL: Crossway, 2021), 96.

my son and I knocked on his door to give him the package, we introduced ourselves. After exchanging pleasantries, we learned that he was a Christian and had gone to school with one of my seminary professors. As I reflected on it later that day, it struck me what a strange time we are living in.

You see, in any other age, my local church would consist of the professing believers that live in my geographical vicinity. I have neighbors in every direction who are Christians, but we aren't a part of the same assembly. Some of them I have never even met. Why not? Because of something called the automobile. Cars, as an extension of human feet, make it possible for us to go remarkably great distances in little time at all. The members of my church are spread out over a 70-mile radius! That would have been impossible before the invention of the car. Now I am doing church with people who live physically distant from me, and it fundamentally alters the nature of our relationship. I see my next-door neighbor on a nearly daily basis, especially in the warmer months. The physical interaction and engagement happen spontaneously when you live near one another. But through technological innovation, the believers I am united together with in a local church don't live in my immediate locality (something altogether unprecedented in church history).

What does that mean for Cornerstone Presbyterian Church? It doesn't mean we should sell our cars. Nor does it mean we should dissolve as a church. Nor does it necessarily mean that we should seek to move closer to one another. But it does mean fellowship and hospitality that may have happened spontaneously in prior eras are now going to require strategic planning and conscious effort.

Technology changes how we fellowship as the church, and what is true of cars is even more true of digital communication mediums. It is not that there is no place for such, but we need to think very carefully about how they fundamentally alter the way we relate. For the believers in the early church were intimately involved in each other's lives in physical and often inconvenient and costly ways (e.g., Acts 2:42–47; 4:32–37). While the predominant focus in *Sacred Streaming?* is on the church's worship, our embodied existence and physical presence is a key component of church fellowship too, and we need to recognize that digital technology poses a great threat to that.

The Church of the Metaverse

If we think that livestream is a threat to the embodied nature of the church's worship and fellowship, the threat will only grow in the days ahead. Livestream as we know it will quickly become a thing of the past just like pagers and VCRs. For Big Tech is determined to create a disembodied universe that appears so real that your senses are deceived into thinking you are embodied in it. They call it the metaverse.

Mark Zuckerberg describes it as "an embodied internet, where instead of just viewing content — you are in it. And you feel present with other people as if you were in other places, having different experiences that you couldn't necessarily do on a 2D app or webpage, like dancing, for example, or different types of fitness." The internet is a disembodied digital environment, but Zuckerberg is intent on making it *feel* otherwise. He recognizes that the current digital devices and platforms we have for communicating go against our bodily nature and do not enable us to be fully present. But he intends to change that. He explains,

> What virtual and augmented reality can do, and what the metaverse broadly is going to help people experience, is a sense of presence that I think is just much more natural in the way that we're made to interact. And I think it will be more comfortable. The interactions that we have will be a lot richer, they'll feel real. In the future, instead of just doing this over a phone call, you'll be able to sit as a hologram on my couch, or I'll be able to sit as a hologram on your couch, and it'll actually feel like we're in the same place, even if we're in different states or hundreds of miles apart.[37]

This is where Silicon Valley desires to go, and you're sorely mistaken if you think they won't seek to take the church with them. Through a VR helmet or a brain chip, you will be able to virtually enter a church building, walk around inside, interact with people, and engage in public worship—all without ever leaving your living room. And according to Zuckerberg, you will "feel present."

But regardless of how you might feel, you won't be present, at least not in the flesh-and-blood way that God created and redeemed you to be present in the sacred assembly. When we begin to reason otherwise, settling for a so-called "technological presence" which will only feel more real with every technological advance, we are in serious danger of being conformed to the body-degrading dualism that is inimical to the church's worship and fellowship. Matthew Lee Anderson warns, "Without an anthropology that affirms not only the body's goodness but also the need for the body's presence within the communal

37. Casey Newton, "Mark in the Metaverse," *The Verge*, July 22, 2021.

life and corporate worship of the church, there is nothing to hold in check the anti-material thrust of modern technological development."[38] Beware, dear reader.

38. Anderson, *Earthen Vessels*, 217.

CHAPTER 4

BEWARE OF AN
INCURABLE APATHY

A N EIGHTEENTH-CENTURY preacher once jotted down a
bright vision of technological advance. His name was
Jonathan Edwards, and he dreamed of a golden era of spiritual
blessing for the church when "they will have better contrivanc-
es for assisting one another through the whole earth, by a more
expedite and easy and safe communication between distant
regions than now." Connecting Christians scattered around
the globe was impractical, and in many cases impossible, in
Edwards' day. But in the future a medium of communication
would unite them speedily, effortlessly, and safely. It would
be comparable to some of the greatest innovations up to that
point. "The invention of the mariner's compass is one thing by
God discovered to the world for that end; and how exceedingly
has that one thing enlarged and facilitated communication!"
For many centuries this handheld contraption with its magne-
tized pointer had guided national and international travelers,

enabling the gospel to spread and connecting Christians that otherwise would have never crossed paths. The compass, however, did not jettison the slow, difficult, and dangerous trek across land and sea; it merely directed it. Edwards envisioned a future when technology would eliminate the need for the trek altogether. "And who can tell but that God will yet make it more perfect; so that there need not be such a tedious voyage in order to hear from the other hemisphere, and so the countries about the poles need no longer to lie hid to us, but the whole earth may be as one community, one body in Christ."[1] Through this mysterious medium Christians on different sides of the hemisphere would be able to hear and see one another. In Edwards' mind this would unite the universal church visibly on earth in a way never known before, and it would be glorious!

One can't help but read Edwards' words penned in 1725 and recognize that exactly three centuries later we are living his dream, except that it is not nearly as bright as he imagined it to be. Digital technology has made it possible to speak to, hear, and even see fellow Christians in other parts of the world with the click of a button. Human innovation has "enlarged and facilitated communication" so that "a tedious voyage" is no longer required, but it hasn't been accompanied by the period of ecclesiastical unity and prosperity which Edwards elsewhere called "the church's latter-day glory."[2]

Given his reading of biblical prophecy, Edwards anticipated a future when the Spirit would be poured out in an ex-

---

1. Jonathan Edwards, *The "Miscellanies": Entry Nos. A–z, Aa–zz, 1–500*, ed. Thomas A. Schafer and Harry S. Stout, vol. 13, *The Works of Jonathan Edwards* (New Haven, CT: Yale, 2002), 369.

2. Jonathan Edwards, *The Great Awakening*, ed. C. C. Goen, vol. 4, *The Works of Jonathan Edwards* (New Haven, CT: Yale, 1972), 358.

traordinary way around the world, resulting in unprecedented spiritual and numerical growth in the church until Christ returned. He did not believe human innovation would accomplish this—only the Spirit-empowered word could! But he did believe technological advance would accompany, encourage, and facilitate this mighty movement of God in the last days by making gospel communication swift, simple, and safe.

There is no denying the fact that radio, telephone, and internet have been used for great good, connecting distant believers and propagating the word to unreached peoples and places. Praise God for this! But these same technologies have facilitated what is arguably one of the greatest dangers the church has ever known, polarizing Christians and promoting a shallow, privatized religion.

If Edwards were living today, would he maintain his technological optimism? Might five minutes on Twitter strike a deathblow to his postmillennial vision altogether?[3]

Yet what Edwards hoped for is in a very real sense God's aim for His people throughout the present age. He desires our spiritual unity and prosperity. This is the reason He assembles His people week after week in His holy presence.

We fleshed out the substance of the church's worship in chapter two, seeing that there is an incongruity between livestream and the inherently public nature of public worship. In chapter three we examined the elements of the church's wor-

---

3. I'm joking, of course. Edwards believed what he did about a future millennial reign of Christ because of his reading of Scripture, not because of human experience. While I believe the Scriptures teach that the entire church age is the millennium wherein Christ's rule is extending mightily, I deeply respect Edwards' rigorous biblical exposition concerning the end times.

ship and the fact that digital technology is a less-than-adequate form through which to carry them out since every element requires bodily engagement and physical proximity. Now we turn to consider the aim of public worship.

What is God's great goal in gathering us in local assemblies on His day? I propose that His aim in calling the church to worship is *an earnest, exclusive, expanding, and eschatological devotion.* This radically Godward consecration is precisely what Edwards hoped for on a global scale. We should too. But we can't assume that modern communication technologies will necessarily facilitate such an end. Livestream may actually hinder and harm it.

Devotion to the Supreme Good

Our God is holy. The root of the Hebrew adjective means "to cut" or "to separate." It involves a divide as the holy object is cut off and separated from other objects. Theologians have traditionally understood God's holiness to refer to His absolute separation from creation and corruption. R.C. Sproul concisely describes this attribute as "transcendent purity."[4] God is transcendent, being uncreated. He is pure, being uncorrupted.

This understanding of holiness is true as far as it goes, but it doesn't go far enough. I'll explain why by way of a question— did God become holy only after He created and the creation rebelled? The obvious answer is *no.* God never *became* holy; He eternally *is* holy. But if holiness is nothing more than His separateness from creation and corruption, then there could be no holiness before there was creation and corruption to be separate from. Before the beginning of time and space, there was

---

4. R. C. Sproul, *The Holiness of God,* 2nd ed. (Carol Stream, IL: Tyndale, 1998), 38.

nothing save the holy God. What then does it mean for God to be holy when there is nothing for Him to be set apart from?

We are given a clue in the objects God sets apart as holy throughout redemptive history. Take, for example, the Sabbath: "Six days shall work be done, but on the seventh day is a Sabbath of solemn rest, a holy convocation. You shall do no work. It is a Sabbath to the LORD in all your dwelling places" (Lev. 23:3). The Sabbath is a day set apart from the six ordinary days of work, but it is no bare separation. It is a separation unto service. It is a day "to the LORD" as God's people cease their ordinary vocations ("solemn rest") to gather with the sacred throng in worship ("holy convocation"). This day is certainly cut off (i.e., division), but it is cut off in order to be entirely given over to God (i.e., devotion).

Devotion, not division, is the essential core of holiness. Though having nothing to be separate from prior to the creation and the fall, God was absolutely devoted to Himself as the supreme good. Hence, Sinclair Ferguson describes this attribute of attributes as "the perfectly pure devotion of each of the three persons to the other two...absolute, permanent, exclusive, pure, irreversible, and fully expressed devotion."[5] Richard Muller calls it God's "sacred self-regard."[6] Geerhardus Vos extols the divine holiness as "that attribute of God by which He seeks and loves Himself as the highest good."[7] More

---

5. Sinclair B. Ferguson, *Devoted to God: Blueprints for Sanctification* (Edinburgh: Banner of Truth, 2016), 2.

6. Richard A. Muller, *The Divine Essence and Attributes,* vol. 3, *Post-Reformation Reformed Dogmatics: The Rise and Development of Reformed Orthodoxy, ca. 1520 to ca. 1725,* 2nd ed. (Grand Rapids: Baker Academic, 2003), 500.

7. Geerhardus Vos, *Reformed Dogmatics*, vol. 1, *Theology Proper*, trans. and ed. Richard B. Gaffin Jr. (Bellingham, WA: Lexham, 2012), 27.

than an absolute division from all that is not Himself, God's holiness is an absolute devotion to Himself.

Covenant history is the story of God seeking after a people who reflect His holy glory. Stamped with the divine image, humans are uniquely set apart from the rest of creation to see, savor, and serve God as the highest good. Through the gospel, God restores us to the image we have fallen from, calling us out of the corrupt world to belong to Him in whole-person consecration: "You shall be holy, for I am holy" (Lev. 11:44; 1 Pet. 1:16).

While all of life is to be devoted to God, public worship on the Lord's Day is the supreme context in which the church as "a holy priesthood" (1 Pet. 2:5) and "a holy nation" (1 Pet. 2:9) expresses its devotion to God. The sacred assembly is also the supreme context in which God purifies the hearts of His people, uniting them in beautiful consecration. From the call to worship to the benediction, God declares to His beloved people, "You are mine!" As the word is read, preached, seen, sung, and prayed, the church is lifted up to Zion to behold God's august holiness which is heaven's preoccupation (Isa. 6:3; Rev. 4:8). Seeing Him by faith, they are transformed to reflect His holiness from one degree of devotion to the next (2 Cor. 3:18). God's holy convocation, on His holy day, in His holy presence, under His holy ordinances, is the ultimate place wherein we simultaneously serve God and are made fit for His service.

Do you realize this is why the church's public gatherings for worship are often referred to as *services*? We assemble to render devoted service to the God who has devoted Himself to us in Jesus Christ, and as we commune with Him through the public means of grace we cannot help but grow in such God-like and God-ward devotion.

The digital screen is arguably a less-than-fitting medium to facilitate this aim. Do you remember the greeting to virtual worshipers that my family heard when visiting a church recently? "If you are watching us on livestream, thank you for coming!" The screen is designed to facilitate watching, not worshiping. It is bent to make us passive spectators, not active servants.

We can, of course, push against this technological bent if we choose to use livestream for religious purposes. But the mode is working against us, not for us, when, for example, a family watches a movie on Friday night for entertainment, only to gather around the same screen on Sunday morning for "church." The form predisposes them to watch the streamed service for entertainment regardless of the content. Rather than assisting us in giving and growing in devotion to God, livestream has the inbred tendency to mold us into religious watchers.

The ill-disposed nature of livestream in the church's service only becomes more apparent when we consider exactly what kind of devotion God wills us to possess and pursue as His gathered people.

An Earnest Devotion

J. C. Ryle lamented in the nineteenth century "the lazy, easy, sleepy Christianity of these latter days."[8] While some Christians stream worship services today because they are unable to physically gather with the saints, others do so out of a deplorable apathy. In far too many instances livestream has become the vehicle of a lazy Christianity that doesn't require getting off the couch, an easy Christianity that involves nothing more

---

8. J. C. Ryle, *Practical Religion: Being plain papers on the daily duties, experience, dangers, and privileges of professing Christians* (Edinburgh: Banner of Truth, 2013), 173.

than a swipe of the finger, and a sleepy Christianity that encourages passive engagement.

Such religious apathy is not worthy of the name Christian. It is the exact opposite of the earnest zeal God wills His people to possess: "a burning desire to please God, to do his will, and to advance his glory in the world in every possible way."[9] The messianic King confessed, "Zeal for your house has consumed me" (Ps. 69:9), and such fiery zeal for God and His house ought to consume the hearts of His followers. "Do not be slothful in zeal, be fervent in spirit, serve the Lord" (Rom. 12:11). Sadly, it is all too easy for our affections for God to cool and for our devotion to be less than earnest.

We don't need livestream for that to happen. Apathy is all too common in the public assembly of the saints. In relation to corporate worship, Octavius Winslow warns,

> This state of secret departure from God may exist in connexion with an outward and rigid observance of the means of grace; and yet there shall be no spiritual use of, or enjoyment in, the means. And this, it may be, is the great lullaby of his soul. Rocked to sleep by a mere formal religion, the believer is beguiled into the delusion that his heart is right, and his soul prosperous in the sight of God.[10]

It is possible for the public singing, praying, reading, and preaching to lull us into a spiritual slumber, deceiving us into thinking we are engaging with God when in reality our religion is one of mere formality and dead externals. You can get up off the couch and assemble physically with the church all in the

9. Ryle, *Practical Religion,* 174.

10. Octavius Winslow, *Personal Declension and Revival of Religion in the Soul* (1841; Edinburgh: Banner of Truth, 2021), 9.

service of a "lazy, easy, sleepy Christianity." So, what's the big deal about livestream?

If apathy can be fostered through the public means of grace when we make the effort to physically assemble, how much more so when those gracious ordinances are mediated through a screen! The potential to grow cold and slothful in our service to God goes through the roof when the church and her worship are digitized. Why?

For one thing, the latest modes of digital communication are not fit to carry weighty messages. I still write hand-written, paper-and-ink love letters to my wife (though not as frequently as I should). It would be much easier to type up my romantic utterances to send in an email or to bake them down into a few lines to send in a text, but the ease of those modes of communication would fundamentally gut the message of its weight. This is, as John Dyer points out, "part of the reason older mediums tend to communicate a deeper sense of meaning and value than newer mediums do."[11] Call me old fashion, but no form of digital communication will ever be able to replicate the gravity of hand-written correspondence or face-to-face communication. Digital streaming technologies have an inherent tendency to gut whatever message is being communicated of its weight. They are simply too easy, speedy, and comfortable to use. Thus, whatever propensity we might have to be lulled into a slumber by public worship is only heightened when we utilize such a lightweight medium.

Which brings us to a related reason livestream poses a threat to our zeal for the Lord—it is incredibly difficult to be attentive when you are not physically present. Research shows

---

11. John Dyer, *From Garden to the City: The Redeeming and Corrupting Power of Technology* (Grand Rapids: Kregel, 2011), 120.

that online students tend to get lower grades and express greater difficulty in learning than in-person students.[12] During the
Covid pandemic, I was forced to stream seminary lectures for
much of my final semester. I often joke that I didn't learn anything from those classes! That may be a bit of an exaggeration,
but only a bit. The fault was not my professors on the screen.
The fault was not the content being conveyed. The fault lay
squarely with the technological medium. Given the oratorical
and communal realties we examined in chapter 3, lectures are
much less engaging through a screen, and distractions are much
more tantalizing. I did things during those livestreamed lectures that I never would have done while sitting in a classroom.
Neil Postman, writing about the phenomenon of watching TV
preachers, explains, "People will eat, talk, go to the bathroom,
do push-ups or any of the things they are accustomed to doing
in the presence of an animated television screen."[13] If I do all
those things when I stream a Netflix show, why not do them
when I stream a lecture or a sermon? Why not check my email,
get my work schedule organized for the coming week, and
make sure Sunday lunch is in the oven all while watching the
church service? I wouldn't have the audacity to do any of those
things when in the public assembly, but when in the privacy
of my own home using this lightweight medium to engage in
worship, why not? If you have ever livestreamed a church ser-

---

12   See, for example, Heppen, J. B., Sorensen, N., Allensworth, E., Wal-
     ters, K., Rickles, J., Taylor, S. S., & Michelman, V., "The Struggle
     to Pass Algebra: Online vs. Face-to-Face Credit Recovery for At-
     Risk Urban Students," *Journal of Research on Educational Effective-
     ness*, 10(2), 272–296.

13. Neil Postman, *Amusing Ourselves to Death: Public Discourage in the Age
     of Show Business* (New York: Penguin, 1985), 119.

vice, you understand what a serious temptation this is, and it is a temptation that is wired into the technological form itself.

Livestream makes it incredibly difficult to maintain and foster the earnest, aggressive, attentive devotion God wills for us to have in His worship. And without that earnestness, we ought not to expect His blessing. For the soul-satisfying riches of communion with God are not obtained through half-hearted, distracted engagement. "He that will not have the *sweat*, must not expect the *sweet* of religion," writes John Flavel.[14] Or to put it positively, *the sweat comes before the sweet.*

I often physically sweat in public worship, especially in the act of preaching. But even if your body is not drenched in perspiration, your soul ought to be. For there is nothing passive about worship. Religious spectators "at ease in Zion" forfeit the sweetness of God's blessing and might even end up tasting His bitter curse (Amos 6:1). We need to be very watchful against the lazy, easy, sleepy disposition toward God and His worship that a technological lightweight like livestream can so easily foster in us.

An Exclusive Devotion

In our exposition of the Decalogue, we saw there is a back-and-forth interplay between the first two commandments. Who we worship determines how we worship, and "how we worship determines who we worship."[15]

---

14. John Flavel, *Saint Indeed*, in *The Works of John Flavel* (Edinburgh: Banner of Truth, 1968), 5:495.

15. J. Ligon Duncan III, "Does God Care How We Worship?" in *Give Praise to God: A Vision for Reforming Worship,* eds. Philip Graham Ryken, Derek W. H. Thomas, and J. Ligon Duncan III (Philipsburg, NJ: Presbyterian & Reformed, 2003), 33.

As the First Commandment makes clear, there is only one object worthy of our affectionate, whole-person exaltation— the Triune God of creation and redemption. "You shall have no other gods before me" (Exod. 20:3). God demands our exclusive devotion.

It was the forfeiture of this single-eyed love to God that plunged the human race into death. Our first parents made an idol of a piece of fruit, believing this edible object was more essential than God and could satisfy their hearts in a way God could not (Gen. 3:1–6). The apostle, writing to a church being infiltrated by false teaching, expressed fear "that as the serpent deceived Eve by his cunning, your thoughts will be led astray from a sincere and pure devotion to Christ" (2 Cor. 11:3). These Christians didn't need to go to the pagan temple to devote themselves to idols; they could do it while assembled as God's temple in worship. That was why Paul asked them earlier in the letter, "What agreement has the temple of God with idols?" (2 Cor. 6:16). Since God had made them His dwelling place, they needed to "cleanse [themselves] from every defilement of body and spirit, bringing holiness to completion in the fear of God" (2 Cor. 7:1). The church must labor to be totally separate from corruption in order to be exclusively devoted to God.

This is God's aim in publicly assembling us—cleansing and consecration that preserves us from going the way of covenant-breaking idolatry like the first Adam (and Israel after him).

Satan and his minions labor tirelessly to pollute the saints' purity of devotion, tempting us with idols of all shapes and sizes. But at the root of them all is the idol of self. The "self-sufficient humanism"[16] that characterizes Western idolatry today is nothing new. For the eating of the forbidden fruit, writes Brian Rosner, involved "several of the tenets of expressive individualism." Rewind to that dark day, and here is what you find Adam and Eve doing: "They reject an authority external to themselves, believe that their existence will improve dramatically as they assert their freedom, and make moral judgments according to personal preference."[17] They made the self the chief end of their existence. It happened in the first earthly temple with the first priests, and it can happen in God's new covenant temple with His new covenant priests.[18]

The forms and mediums we use in worship can easily lead us down self-exalting, idolatrous paths. There are five iterations of the idol of self that we are particularly prone to when we digitize church.

1. The Idol of Self-Preservation

When the Covid pandemic hit, we were told it was not safe to physically gather. Livestream was embraced as the safe alternative to the public assembly. God's people ought not to be reck-

---

16. Charles Taylor, *A Secular Age* (Cambridge, MA: Harvard University Press, 2007), 18.

17. Brain Rosner, *How to Find Yourself: Why Looking Inward is Not the Answer* (Wheaton, IL: Crossway, 2022), 62.

18. Given the sacramental nature of the tree of the knowledge of good and evil, we could even say that the first human instance of idolatry came about through the self-serving misuse and abuse of God's covenant sign and seal. The sacraments, along with all the other prescribed elements of new covenant worship, can become the means by which idolatry is propagated today.

less (Prov. 14:16), but it is helpful to remember that nowhere in the Bible are we commanded, "Be safe."

In *The Coddling of the American Mind,* Greg Lukianoff and Jonathan Haidt show how in recent generations safety has become the all-consuming concern. They call it "safety-ism" which they define as "a culture or belief system in which safety has become a sacred value, which means that people become unwilling to make tradeoffs demanded by other practical and moral concerns. 'Safety' trumps everything else, no matter how unlikely or trivial the potential danger."[19] Our culture of self-safety has latched on to livestream, and many professing Christians embrace it as a legitimate ecclesiastical medium simply because it is safer. There is no fear of deadly germs spreading, murderous shootings transpiring, or uncomfortable conversations unfolding.

Staying home may be safer, but the culture God wills His church to have is not one of self-safety, but self-sacrifice. There may be a time in the midst of plague or natural disaster when it would be reckless and unwise for the church to gather. I'm not denying that. But let us not forget that following Jesus is costly, sometimes even deadly. The call to assemble is coupled with the call to take up our crosses in self-denying discipleship (Lk. 14:27).

Our persecuted brothers and sisters understand this. At the time of writing, Pastor Wang Yi is suffering in a Chinese prison. One of the main reasons for his detainment is that he continued to call his congregation to assemble for Lord's Day worship in spite of the threats of the Communist party. After

19. Greg Lukianoff and Jonathan Haidt, *The Coddling of the American Mind: How Good Intentions and Bad Ideas are Setting Up a Generation for Failure* (New York: Penguin, 2018), 27.

he and a number of his church members were arrested in 2018 he wrote, "Under no circumstances will we stop or give up on gathering publicly, especially the corporate worship of believers on Sunday....I will not cooperate with the police banning, shutting down, dissolving, or sealing up the church and its gathering. I will not stop convening, hosting and participating in the church's public worship, until the police seizes my personal freedom by force."[20] Yi joins a host of faithful believers throughout the ages who were willing to sacrifice themselves to uphold God's will for His church and His worship.

The dangers we face for assembling in the West are trivial in comparison, and yet how quick we are to exalt public safety over public assembly, often in the name of love. Livestreaming church gives us the illusion of not forsaking the assembly while all the while our hearts might be serving the idol of self-preservation instead of the true and living God. When church is no longer costly, we need to question whether it is Christian.

2. The Idol of Self-Indulgence

We've seen how livestream tends to make us watchers in search of entertainment rather than worshipers in search of God. This kind of spectator religion is not new. Sinclair Ferguson notes, "The sixteenth-century Reformers shared a deep, underlying concern that late medieval worship had become a kind of spectator event. The congregation was largely passive. 'Worshipers,' if they could be thus described, were essentially observers of the drama of the Mass, and listeners to the words of the choir."[21]

---

20. Wang Yi, *Faithful Disobedience: Writing on Church and State from a Chinese House Church Movement*, eds. Hannah Nation and J. D. Tseng (Downers Grove, IL: IVP Academic, 2022), 216.

21. Sinclair Ferguson, *Reformation Worship* (Greensboro, NC: New Growth Press, 2018), xvii.

One of the main goals of the European Reformation was to turn religious watchers into religious worshipers, and part of the strategy to get there was a radical reform of the forms by which the elements were carried out. If a form is promoting spectator religion, it ought not to be tolerated in the church.

Even in Reformed circles where entertainment in worship is eschewed, we can still fall prey to "edutainment". If entertainment utilizes music, lights, cinematography, humor, and drama to give us affectional pleasure, then edutainment uses doctrinally-precise, attention-grabbing teaching and preaching to give us intellectual pleasure. We can gather physically for the purpose of being edutained, but virtual "church" makes Reformed Christians particularly prone to this. For it enables us to access the ministries of our favorite preachers who scratch us precisely where we doctrinally itch. It enables us to watch Bible teachers who are far more engaging than any local pastor is. It easily turns us into religious consumers who find the preaching that most pleases us so that worship becomes about indulging our desires rather than exalting God.

Solomon's warning certainly applies here: "Whoever isolates himself seeks his own desire; he breaks out against all sound judgment" (Prov. 18:1). If the preacher says something we don't like, we shut him off and search for a new one. If the sermon is on a text or topic that doesn't interest us, we switch YouTube channels. If the messenger is a bit longwinded, we conveniently mute him or adjust him to 1.75x speed. The sermon becomes just another TED talk designed to edutain us in our busy, self-focused lives.

3. The Idol of Self-Service

Even if we don't stream a service to be entertained or edutained, experiencing the public assembly through a private screen promotes the idea that the worship service exists to serve our personal spiritual needs. There is, of course, truth to this. God gathers each individual believer so that they might "go from strength to strength [as] each one appears before God in Zion" (Ps. 84:7). He graciously serves us in an intimate and personal way in public worship.

But the apostle Paul tells us that Christ's aim through the church's public teaching ministry has a focus beyond our individual selves. Our Lord gives teachers and preachers "to equip the saints for the work of ministry, for building up the body of Christ" (Eph. 4:12). Every saint (i.e., holy one) has unique gifts to exercise toward fellow believers (Eph. 4:7). It is the Spirit-empowered preaching of the word that equips them to do so, resulting in the church's spiritual and numerical growth (Eph. 4:13–16). In the public assembly, individual Christians are to "strive to excel in building up the church" (1 Cor. 14:12). The apostle "challenges the common assumption that church services should simply be designed to facilitate a private communion with God, either by spiritual exercises or ritual. He envisages that believers will come together for the benefit of one another, drawing on the resources of Christ for spiritual growth by the giving and receiving of Spirit-inspired ministries."[22] But when the only connection we have to fellow believers is through a shared server, it is very difficult to respond to the word in others-oriented service.

---

22. David Peterson, *Engaging with God: A Biblical Theology of Worship* (Downers Grove, IL: IVP Academic, 1992), 212.

While we might not watch the service to indulge our flesh, the lack of public gathering prevents us from genuine interaction with others. It robs us of the opportunities to serve that organically arise when physically together. I shut off my screen at the end of the service and am left to me, myself, and I.

Livestream molds us to approach Christianity in a privatized manner. While the gospel "shifts the centre of gravity from the individual to the church,"[23] ecclesiastical streaming shifts the centre of gravity from the church to the individual. In so doing, it not only robs us of opportunities to build up fellow believers with our God-given gifts, but it also stunts our spiritual growth. For it is through selfless, sacrificial service to others that God delivers us from the destructive tendencies of our self-serving pride and conforms us to the humble Christ.

4. The Idol of Self-Sufficiency

If Christianity is a me-and-Jesus religion, then I really don't need anyone except for Jesus. But using the Scriptural analogy of a body, if I am a finger, I need a hand, an arm, a shoulder, and a neck if I would be connected to the head (1 Cor. 12:12–27). Every member of the body is dependent upon every other member of the body. We need each other. Those who forsake the assembly of the saints communicate otherwise. Their contentment with the impersonal, disembodied digitization of church reveals the conviction that they don't need the rest of the body (since that body is necessarily embodied!).[24]

---

23. J. K. S. Reid, *Our Life in Christ* (Philadelphia: Westminster Press, 1963), 94.

24. As has been said throughout the book, there are some who livestream services because they simply have no way of physically assembling. These comments apply to the homebody, not the homebound. But

Is it by accident that the exhortation against neglecting the ecclesiastical gathering (Heb. 10:25) comes in the context of what is arguably the greatest warning against apostasy in the Bible (Heb. 10:26–31)? Certainly not. We need the scriptural exhortations of fellow brothers and sisters lest we drift away (Heb. 3:12–13). We need the faithful oversight of leaders who watch out for our souls lest we stray (Heb. 13:17). But when our relation to the church never goes beyond the screen of a personal electronic device, such interpersonal encouragement and accountability become a practical impossibility.

Out of God's very good creation, there was only one thing that was not good—Adam's lack of human companionship (Gen. 2:18). "Two are better than one," writes Solomon. Why? "For if they fall, one will lift up his fellow. But woe to him who is alone when he falls and has not another to lift him up! Again, if two lie together, they keep warm, but how can one keep warm alone? And though a man might prevail against one who is alone, two will withstand him—a threefold cord is not quickly broken" (Eccl. 4:9–12). Isolation leaves us incredibly vulnerable. How many professing Christians spiritually fall and never get back up because they have deserted the very ones with the ability to draw them out of the sinful pit? How many professing Christians have grown cold toward God and His word due to a prolonged failure to stoke their hearts by the fires of Christian fellowship? Samuel Ward asks, "Such as forsake the best fellowship, and wax strange to the holy assemblies (as now the manner of many is), how can they but take cold?

there is a subtle message being communicating by the privatizing of the public through this technological medium, and even the homebound who prize the members of the local church need to beware of the subtle ways it might be shaping them.

Can one coal alone keep itself glowing?"[25] We need more than digitally-transmitted sermons on the Lord's Day; we need each other in flesh-and-blood human engagement. Self-sufficiency is a real danger, and it is one that the form of livestream tends to encourage even when it is used with good intentions.

5. The Idol of Self-Righteousness

Despite what expressive individualism wants us to think, identity formation is a community endeavor. David Jopling puts it well: "Persons come to know themselves in being known by persons other than themselves."[26] The Christian who is a stranger to the church is to some degree a stranger to himself. When living in isolation from fellow believers, we are prone to a very skewed self-perception.

For most of us that distorted identity veers in the direction of self-righteousness. In our pride, we tend to minimize our sins and faults while maximizing the sins and faults of others. Part of God's purpose in the public assembly is to bring our sin to light. He does this through the teaching and preaching of His word. But often God chooses to expose us through personal engagement with leaders and fellow believers who know us intimately.

Some of the most defining moments of my Christian life have been when a brother or sister has helped me to see sin in my life that I couldn't see myself. "Faithful are the wounds of a friend" (Prov. 27:6). But such friendship would have never happened apart from our gathering together as members of the

---

25. Samuel Ward, "A Coal from the Altar to Kindle the Holy Fire of Zeal," in *Sermons and Treatises* (Edinburgh: Banner of Truth, 1996), 83.

26. David A. Jopling, *Self-Knowledge and the Self* (New York: Routledge, 2000), 166.

same local church. And without intimate friendship, wounds are bound to be misdirected and misunderstood. For admonition and rebuke ordinarily require a loving relationship in which both parties know and are known by each other.

I'm certain I would be a much more arrogant man with a much shallower view of my own sin apart from the embodied fellowship of the church. That is not only because of the constructive critique of fellow Christians, but also because it is far more difficult to put on a religious façade when you are living in intimate proximity to others. Church life is messy, and it serves to expose the remaining corruption in our hearts. Just this week I've been convicted of sinful irritation, sinful envy, and sinful insecurity, and in each case it has been interactions with fellow saints that God has used to lay bare my twisted heart.

It's easy to create a perfect persona when you are nothing more than digital pixels on a screen (just look at the average social media profile), but when you brush shoulders with other saved sinners in the context of the church, the real you is bound to come out. This is a mercy from God because so long as we are self-righteous, we will never cast ourselves upon the Christ who alone can deliver us from the penalty and power of sin.

Keep Yourself from Idols

We must be on constant guard against idols (1 Jn. 5:21), and when it comes to digitizing church, we must be particularly watchful against the idols of self-preservation, self-indulgence, self-service, self-sufficiency, and self-righteousness. It is certainly possible to use this technological medium in the church without succumbing to such, but it is very difficult. Just as false teaching distorted the pulpit ministry of the church at Corinth, threatening their "pure devotion to Christ" (2 Cor. 11:3), so too inappropriate forms in public worship can strike a fatal blow to the same. It is frighteningly possible to serve self-exalting idols in the name of Christian worship, and livestream makes it all the easier for that possibility to become reality.

An Expanding Devotion

God's great goal in history is to extend the church's earnest and exclusive devotion until the whole earth is filled with worshipers. That is why His first recorded command given to the first humans was, "Be fruitful and multiply and fill the earth and subdue it" (Gen. 1:28). As the image of God, Adam and Eve were to procreate so that their offspring of priestly kings might "widen the boundaries of the Garden in ever increasing circles by extending the order of the garden sanctuary into the inhospitable outer spaces."[27] Humanity was to multiply so that the entire earth might be filled with God's presence and worship. What the first Adam failed to do by his representative idolatry against God, the second Adam obtained by His representative devotion to God (Rom. 5:17). Through Jesus Christ, we are reconciled to God and restored to His image (2 Cor. 5:18–21; Eph. 4:24), enabling us to serve in God's presence as priests and to expand the borders of His rule as kings until "all the ends of the earth fear him" (Ps. 67:7; cf. 1 Pet. 2:4–12; Matt. 28:18–20).

If the church's devotion to God is not expanding, resulting in a growing temple through the multiplication of God-exalting worshipers, then it falls short of what it ought to be. We

---

27. G. K. Beale, *The Temple and the Church's Mission: A Biblical Theology of the Dwelling Place of God*, New Studies in Biblical Theology 17, ed. D. A. Carson (Downers Grove, IL: InterVarsity, 2004), 85.

witness to our children and to a lost world that the Triune God is worthy, calling them to join us in glorifying and enjoying Him through the gospel.

One of the common arguments in favor of livestreaming the church's worship is that it serves this mission by enabling people who would never step foot into our assemblies to encounter them digitally. Matt Peeples, writing in favor of ecclesiastical streaming, asks, "In a culture that's becoming increasingly post-Christian, is it wise to limit the places we're sharing the gospel?"[28] It is intended as a rhetorical question, for no Christian wants to limit the spread of the gospel. In actuality, wisdom would have us limit what mediums we use to expand God's worship. Rather than "getting the word out by every means possible," as Peeples advocates,[29] we ought to get the word out by the best means possible, wisely considering how a certain medium might shape the message and the worshipers it creates.

To a Thousand Generations

Few texts call us to earnest and exclusive devotion like *the Shema:* "Hear, O Israel: The LORD our God, the LORD is one. You shall love the LORD your God with all your heart and with all your soul and with all your might" (Deut. 6:4–5). Our singular God calls us to love Him with everything that is within us by keeping His words in our hearts (v. 6). But this is no individualistic endeavor, for God desires such all-consuming devotion to be passed on from generation to generation: "You shall teach them diligently to your children, and shall talk of

---

28. Matt Peeples, "Why Our Church Will Keep Livestream," *The Gospel Coalition,* May 27, 2021.

29. Peeples, "Why Our Church Will Keep Livestream."

them when you sit in your house, and when you walk by the way, and when you lie down, and when you rise" (v. 7). The present generation of believers is responsible to do everything in their power to propagate the worship of God "to a thousand generations" (Deut. 7:9). This was the goal of God's original commission to Adam (Gen. 1:28), and it continues to be so in the church today (Eph. 6:4). "The evangelization and nurture of the church's children," writes Joel Beeke, "is the greatest and most fruitful means of church growth there ever was."[30]

One of my chief concerns regarding the proliferation of livestream in the church is how it will affect future generations. Wisdom would have us consider the ramifications of our present choices for those who will come after us. Thanks to an ever-growing body of secular research, it's actually not hard to discern how the church's adoption of livestreaming technology is likely to affect its young people.

Jonathan Haidt describes Gen Z (those born after 1995) as "drifting through multiple disembodied networks" thanks to the glowing rectangular boxes in their pockets.[31] Jean Twenge chronicles how this has resulted in the rapid decline of in-person interaction among the generation she aptly labels "iGen." Twenge summarizes the data, "Time spent with friends in person has been replaced by time spent with friends (and virtual friends) online."[32] The past 15 years has been an experiment to

---

30. Joel R. Beeke, *Parenting By God's Promises: How to Raise Children in the Covenant of Grace* (Sanford, FL: Reformation Trust, 2011), 27.

31. Jonathan Haidt, *The Anxious Generation: How the Great Rewiring of Childhood Is Causing an Epidemic of Mental Illness* (New York: Penguin, 2024), 195.

32. Jean M. Twenge, *iGen: Why Today's Super-Connected Kids Are Growing Up Less Rebellious, More Tolerant, Less Happy—and Completely Unpre-*

see what would happen to kids who grew up with unfettered access to the internet and the ability to connect with anyone at any time in any place through social media and online gaming. It has proven to be disastrous, leading to an unprecedented spike in anxiety, depression, self-harm, and suicide.[33] All of this ought to be deeply troubling to us, but one conclusion drawn by Haidt about Gen Z is particularly haunting: "They are less able than any generation in history to put down roots in real-world *communities* populated by known individuals who will still be there a year later."[34]

To the degree this is true, it equals the death of the church. For what is the church if it is not a deeply-rooted community of real people intimately known by God and one another? You don't have to look hard to find manifestations of professing Christians among Gen Z exchanging real church for digital counterfeits. For example, Roblox, a popular gaming platform, has given rise to "The Robloxian Christians" who describe themselves as "a youth-led online church….that defies geographical and demographic borders faced by other physical churches."[35] Founded by an eleven-year-old boy, this "church" enables you to design an avatar through which you virtually attend services and fellowship with other avatars. What might seem bizarre to a Millennial like me is deemed normal by the average Gen Z-er shaped to view church as one of many "disembodied networks" they drift through in a given week in exchange for "real-world communities."

---

pared for Adulthood (New York: Atria, 2017), 73, emphasis original.

33.   Haidt, *The Anxious Generation,* 21–45.

34.   Haidt, *The Anxious Generation,* 194.

35.   https://www.therobloxianchristians.org.

You might tune into your church's YouTube channel only when sickness prevents your family from gathering in-person, seeing virtual church as a less-than-ideal but better-than-nothing alternative. But how might your use of livestream be shaping your kids who are a part of a generation that is on the verge of forsaking in-person everything? If in-person is optional, Gen Z-ers are likely to opt out.

As a father, I sense the need to take extreme measures to communicate clearly to my sons that there is no replacement for the flesh-and-blood assembly of embodied saints. That means livestreaming a worship service is simply not an option in the Thompson house. Period. As for me and my house, we will get off our rears and assemble to serve the Lord! And in the rare instances when we are providentially unable to do so, we will sit on our rears and lament our absence.

I fear that many Christian pastors and parents will look back in thirty years and grieve the fact that their children have grown up to have no real attachment to the church. For once the church is streamable it becomes lightweight and dispensable. Livestreamed worship may be novel for older generations, but it is normal for Gen Z, and only time will tell how they are affected by this widescale religious experiment of digitizing church.

If we take seriously our calling to mold our children and grandchildren into God-exalting worshipers, we must take pains to not only teach them about the church and its worship but to give them every opportunity to experience the sweetness of the church and its worship in all of its embodied physicality and geographical proximity. For, as Al Mohler warns, "parents can hardly claim shock when their kids grow up and leave what they have never really known."[36]

Perhaps the greatest threat of the current livestreaming phenomenon in the church is not how it is making us spiritually numb but how it is paving the way for an incurable epidemic of Godward apathy in the coming generations. Assuming for the sake of argument that livestream is permissible in God's worship, is it wise to use this technological medium given what we know about Gen Z?

That is a rhetorical question.

To the End of the Earth

God wills the church's devotion to spread not only from generation to generation, but also "to the end of the earth" (Is. 52:10). Christ sends His church out on mission to bring the gospel to lost friends, neighbors, co-workers, and classmates. On the surface, livestream might appear to be a great tool to assist us in this endeavor. For many of our acquaintances who would never step foot in a church building might be willing to stream a church service where the good news of Jesus Christ is clearly proclaimed. Furthermore, making our services available on the world wide web gives us the potential to reach the whole wide world. It is certainly a way of getting the gospel to those who might not hear it otherwise.

But the church's mission is not merely to publish the good news. Christ calls us to share the gospel unto the formation

---

36. R. Albert Mohler Jr., *The Gathering Storm: Secularism, Culture, and the Church* (Nashville: Nelson, 2020), 141. Mohler provides three key ways we must reach the next generation with the gospel. Here is the first: "Christian parents must view church as the highest and utmost priority for their family's weekly schedule" (140).

of disciples who are brought into the church through the waters of baptism and are built up through the church's ministry (Matt. 28:18–20). "The mission of the church," explains Kevin DeYoung and Greg Gilbert, "is to go into the world and make disciples by declaring the gospel of Jesus Christ in the power of the Spirit and gathering these disciples into churches, that they might worship the Lord and obey his commands now and in eternity to the glory of God the Father."[37] But when we publish the gospel by livestreaming church services, the unnamed and unknown disciples we might be making are arguably being trained from the outset to perceive the gathered church as unnecessary and inconvenient. Our zeal to spread the gospel could actually be "contributing to the dechurching of America" as we "make it easier to stay disconnected from the body of Christ."[38] While many forms of digital technology can and should be used by the church to seek and save the lost, we must avoid all technological mediums that would foster or facilitate a church-less discipleship.

The prophets spoke of the last days as a time when the nations would flock to God's temple mountain to worship and be taught by Him (e.g., Isa. 2:2–3; 52:7–10; 60:1–3). This is precisely what happened after the Spirit was poured out on the day of Pentecost. The gospel went out into the world through

---

37. Keven DeYoung and Greg Gilbert, *What Is the Mission of the Church? Making Sense of Social Justice, Shalom, and the Great Commission* (Wheaton, IL: Crossway, 2011), 62. J. Gresham Machen puts it simply, "When, according to Christian belief, lost souls are saved, the saved ones become united to the Christian Church" [*Christianity & Liberalism,* 2nd ed. (Grand Rapids: Eerdmans, 2009), 133].

38. Jim Davis and Skyler Flowers, "Why Our Church Will Unplug from Streaming," *The Gospel Coalition,* May 27, 2021.

the church, and those saved through it were drawn into the worshiping community of the local congregation (Acts 2–28). Christ continues to do this work today as He builds up the assembled saints through their worship and fellowship, and then sends them forth to bear witness to Him in the world.

At times salvation comes to the lost, not by the church going to them, but by them coming into the church. Invited by a friend or drawn by curiosity or custom, an unbeliever attends public worship only to find that "the secrets of his heart are disclosed, and so, falling on his face, he will worship God and declare that God is really among you" (1 Cor. 14:25).[39]

Either way, the mission of the church results in disciples who express their newfound devotion to God by formally connecting to the church and engaging in her worship. The individualistic and privatized bent of livestream makes it difficult to see how it could be a fitting form to achieve this grand aim.

An Eschatological Devotion

Whether in the first temple of Eden, the tabernacle-temple of old covenant Israel, or the temple of the new covenant church, God's earthly dwelling place and its worship has always been designed to reflect His heavenly dwelling place and its worship. When God's people worship Him in His chosen place on earth, they are participating in and preparing for the worship of heaven (Heb. 12:22–24). Public worship is "a foreshadow-

---

39. Commenting on this verse, David Peterson notes that "as the church is edified intensively—being strengthened, consolidated, and preserved as the community of God's people—it may also be edified extensively—being enlarged by the conversion of those who may be visiting or invited by Christian friends" (*Engaging with God,* 211). Churches should pray fervently for this extensive edification!

ing and rehearsal for that time when we shall worship in the heavenly sanctuary, with the angelic choirs."[40]

In this way, the church's devotion in this present age has a necessarily eschatological character.[41] In the worshiping assembly we are given a foretaste of the new earth of which we are told, "No longer will there be anything accursed, but the throne of God and of the Lamb will be in it, and his servants will worship him" (Rev. 22:3). In that place the glorified saints will affectionately exalt God with an untiring earnestness and uncompromising exclusivity. "In short," writes Anthony Hoekema, "existence on the new earth will be marked by perfect knowledge of God, perfect enjoyment of God, and perfect service to God."[42] That will be true in every creature and every place, for such Godward devotion will have expanded to encompass the entire cosmos.

Like an excited couple going through the steps of the ceremony at their wedding rehearsal, so too the church on earth eagerly gathers on the Sabbath to rehearse for their eternal Sabbath rest when they will worship, no longer by faith, but by sight. Most would agree that it would be undesirable, impractical, and even inappropriate for a wedding rehearsal to be carried out on Zoom due to the necessity of physical presence and proximity at a wedding. How much more so is this the case as the people of God rehearse for their eternal marriage with Christ!

---

40. Allen P. Ross, *Recalling the Hope of Glory: Biblical Worship from the Garden to the New Creation* (Grand Rapids: Kregel, 2006), 72.

41. I'm using the adjective *eschatological* in its classical sense to refer to the last things, in particular here the eternal state of heaven.

42. Anthony A. Hoekema, *The Bible and the Future* (Grand Rapids: Eerdmans, 1979), 286.

Our Lord did not come from heaven to save avatars or pixels on a screen, nor will He wed Himself to lone individuals. His great aim in redemption is perfect communion with a physical assembly on a physical earth through physical proximity as the physically incarnate Son of God. There will be no digital space or virtual presence in the world to come. The great end of history is a fully-embodied congregation rendering everlasting worship to the incorporeal God who has become mysteriously and unendingly embodied in Christ.

In what meaningful sense can the church's worship foreshadow the worship of the new earth when "gathering" is virtual and "proximity" is digital?

Livestream is a less-than-adequate medium to rehearse for the glory to come, and as such it has the great potential to dull our spiritual senses to that glory.

"What Do You Think, Pastor Jonathan?"

I wish Edwards were alive today. It would be fascinating to set before him his 300-year-old vision of a swift, simple, and safe medium of communication facilitating the church's latter-day glory now that such mediums are realities. How would he respond to this entry from his *Miscellanies* (his notebooks of theological reflections)? Would he tear it out with a chuckle, crumble it up, and exclaim, "What a bunch of naïve rubbish!"? Or would he kindly dismiss me from his study in order to transform his concise paragraph into an extended treatise defending the redemptive use of communication technologies like livestream in the church?

I honestly don't know, and I wish I did. For perhaps I am missing something. Is it possible that my arguments against ecclesiastical livestream are similar to the unfair disdain of the

Great Awakening that Edwards refuted with incredible nuance and care in the final years of his ministry? I am certainly open to the possibility. But I have yet to find a substantial, theologically-grounded, pastoral defense and commendation of sacred streaming. It seems that the majority of church leaders have assumed the medium is neutral and all that matters is the message. So, if the message is the gospel, then the swifter, simpler, and safer it can be communicated, the better! But hopefully you are convinced by this point that such thinking, to put it nicely, is a bunch of naïve rubbish.

If God's aim for public worship is an earnest, exclusive, expanding, and eschatological devotion, then we need both a medium and a message that serves this end. The technology we use to engage in God's worship is no inconsequential matter. For as Tony Reinke cautions, "Misused human innovation makes us spiritually numb."[43] The message could be sound and the prescribed elements of prayer, praise, and preaching could be wielded, but if an inappropriate technological form is being used, all of it could serve as a lullaby to the church's collective soul.

My contention in this chapter is that livestreaming the church's worship and fellowship has the inbred potential to slowly and subtly spread a religious apathy of intergenerational and international proportions. For it is hard to see how this medium is not inimical to God's aim for assembling His people, an aim that will be consummately realized in the new earth.

---

43. Tony Reinke, *God, Technology, and the Christian Life* (Wheaton, IL: Crossway, 2022), 190.

WHAT NOW?

R EVERSING COURSE IS PAINFUL. My wife and I recently experienced this in our parenting journey. We had noticed a growing disrespect of parental authority in our two oldest boys. It was nothing out of control, but it was significant enough for us to seek counsel from godly mentors. We became convicted that the problem stemmed from a neglect of the rod. It's not that we weren't disciplining our boys, but when they had reached the age of seven or so, we had traded the rod for other forms of disciplinary action.

God commends the rod (Prov. 13:24; 22:15; 23:13–14; 29:15), but nowhere are parents explicitly commanded to use this form of discipline, nor does God spell out the precise age brackets in which it is appropriate. Tessa and I were not violating a divine precept, nor was our decision to trade-in the rod at age seven disdaining divinely-revealed parameters. But through the wise counsel of others and prayerful reflection upon our family life in the light of God's word, we became convicted that we had made the trade too early. It was not flagrant disobedience, but it was an unwise decision that was doing our children a significant disservice.

We needed to reverse course, and it was not easy. It was difficult sitting our boys down to tell them that we had erred as their parents, and it was challenging to reinstate the rod in those first days. But reverting to this old, time-tested form of discipline has proven to be eminently effective and fruitful.

As parents, few things have proven more perplexing than helping our boys rightly relate to the ever-evolving technological landscape of the late-modern world. Screen time. Video games. Smart phones. Social media. Virtual reality. Artificial intelligence. What is a parent to do? How can parents be exemplary in their technology use and equip their kids to grow to be wise adults who navigate all things digital unto the glory of God? It's not a simple question, and it doesn't have a simple answer. In fact, like the rod, every family will approach technology differently, and thoughtful parents will certainly find themselves at times needing to reverse their family's technological course. The Thompson house has, and I anticipate we will in the future.

I shared in the introduction how the church I pastor reversed course in 2023 by no longer making our services available to the public via livestream. It was a difficult decision, and it was not a decision every member of our church was happy with at the time. But I am more convinced than ever that it was the right decision for us.

No, I can't point you to a proof text that says, "Thou shall not livestream church," any more than I can point you to a proof text that says, "Thou shall not give your twelve-year-old daughter her own smartphone and Instagram account." I would never want to bind your conscience beyond what God has said. But the cumulative force of the biblical, theological, ethical, and even sociological arguments in this book are certainly reason for pause, and they may be reason for you and your church to reverse course. That is never easy, especially post-pandemic when digitizing the sacred has become normative.

When people hear our church doesn't offer livestream, the typical response is a perplexed "Why not?" We forget how novel and strange ecclesiastical streaming actually is and the fact that, until relatively recently, if someone said, "I went to church," without actually bodily assembling with the saints, they would have been the ones met with wide-eyed confusion. Though ecclesiastical streaming is the norm today that should not deter you or your church from potentially pulling the plug. After all, plenty of things are normal today that you should not embrace.[1]

Nuance Needed

It is difficult to swim against the stream (no pun intended!), but unlike the normalization of something like homosexuality, which is a clear-cut, black-and-white moral issue, I can't tell you that you must swim against the stream of the digitized church in every instance. It's not that simple.

One of the tensions I've felt in writing this book is the categorical difference between a virtual worshiper who is able to physically assemble and one who is unable to physically assemble. The homebody chooses the digital as a better option than the physical, while the homebound chooses the digital as the next-best option to the physical. Those two choices are worlds apart, and I would never want to simplistically lump them into the same category. Reversing course is obviously

---

1. A most serious example is same-sex "marriage," while a less serious example is high-fructose corn syrup.

needed for the homebody, but it is less clear whether that is the case with the homebound.

Furthermore, there is significant nuance to how a church may or may not use livestreaming technology. My church pulled the plug on our public livestream, but we actually still livestream our services. That stream, however, is private, and it is only accessible in our church's nursery. Some might accuse Cornerstone's elders of being woefully inconsistent at this point, but I don't believe we are. The use of livestream in this instance is not communicating that public worship can be enjoyed from the privacy of one's home as if the assembly is inessential. For those in the nursery have physically assembled to engage with Christ and His people. As far as we can discern, utilizing this technological medium in the nursery is also not promoting self-exalting idolatry, threatening to undermine God's aim in public worship. For the ladies working in the nursery have come to pour themselves out in service to other families in the church. The nursing mothers who use the room have gathered with the saints despite the great inconveniences and sleep-depravation that accompany having a newborn. If anything, these women are the epitome of selfless, Godward devotion!

A number of our sister churches have a nursery directly behind their main meeting room separated by a wall of soundproof glass. The audio of the worship service is transmitted into the room via electronic speaker so that those with wailing infants and rambunctious toddlers can hear the service even as they behold it through the glass, all without distracting other worshipers. Our building's floorplan does not allow for such. But though our nursery is far from the room we assemble in for worship, we view livestream as analogous to the soundproof glass and speaker (which are both forms of human technology, by the way).[2] While the screen is not an ideal medium through which to engage in the elements of worship and poses potential dangers (as was argued in chapter 3), we believe streaming services in the nursery is a way to encourage the assembled devotion of the many servant-hearted women in our congregation. Our theology is in the driver's seat, and in this particular instance it leads us to the pastoral use of livestreaming technology.

I share that not to legislate a livestream feed in church nurseries (your church may not even have a nursery), but to show the nuanced approach needed when considering a technological medium like livestream. The digital optimist universally embraces the digital without much thought. The digital pessimist universally rejects the digital without much thought. But the digital minimalist thinks deeply about the digital, utilizing it only when it strongly supports his convictions and mission.

The beautiful thing about the philosophy of digital minimalism, as presented in the introduction, is that no two digital minimalists are identical. One church might make a private livestream available for shut-ins, while another may not. One shut-in might connect to a church's livestream in good conscience, while another may not. The goal of this book is not to provide a universally binding list of dos and don'ts. The goal is to assist you in thinking critically about these matters by helping you to understand the dangers at play in digitizing church. At the end of the day, you must work out the potential application of those threats in your own peculiar context.

---

2. That is not to say there is no difference between really beholding something through the glass of window and virtually beholding something through the glass of a screen.

As we draw things to a close, I want to help you to do that by utilizing T. David Gordon's five biblical models for approaching ethical choices. At the risk of over-simplifying matters, these five paradigms could be summarized in five questions, each of which are important for us to ask in the various ecclesiastical circumstances in which a choice needs to be made whether or not to use livestream:

1. The imitation model: "Which choice will best encourage and enable God's image-bearers to reflect Him?"
2. The law model: "Which choice will best accord with God's commandments summarized in the Decalogue?"
3. The wisdom model: "Which choice will lead to the best outcomes in the near and distant future?"
4. The communion model: "Which choice will best foster fellowship with God?"
5. The warfare model: "Which choice will best counter the devices of Satan, offensively or defensively?"[3]

Let's briefly consider digital church through the lens of each of these ethical paradigms.

The Imitation Model

God created humanity in His image to reflect Him as His representatives (Gen. 1:26–27). Without becoming divine, we are called to be God-like. One of the ways we do this is by imitating the divinely prescribed pattern of working six days and resting one. The Sabbath, codified in the fourth commandment, is a creation ordinance that "would advise [man] that his life

---

3. T. David Gordon, *Choose Better: Five Biblical Models for Making Ethical Decisions* (Philipsburg, NJ: Presbyterian & Reformed, 2024).

in this world was patterned after the divine example."[4] Upon the basis of His creational rest, God calls His image to rest, not by sleeping in and devoting one day a week to leisure, but by entering into His holy presence with His people in worship. The Sabbath is a day holy unto the Lord (Ex. 20:8–11). It is set apart from the other days to be entirely devoted to Him, and He wills for that devotion to come to expression in the public assembly (Lev. 23:3). When we assemble in local churches for worship on Sunday, we are reflecting God.

While God does not have a body, He calls us to reflect Him with our bodies. For since "the image of God extends to man in his entirety," reasons Anthony Hoekema, "we must also include the body as part of the image."[5] So the fourth commandment which calls us to imitate God's work-rest pattern is a call to whole-person ceasing from work and whole-person assembling for worship with God's holy ones.

As perfect man, Christ is the perfect image and becomes the pattern we are called to reflect (1 Jn. 2:6). "Be imitators of me," beckons the apostle, "as I am of Christ" (1 Cor. 11:1). A closer look at the life of Christ reveals that from the youngest age He was devoted to public worship. He learned this devotion from His parents who dedicated Him to God in the temple according to the law (Lk. 2:22) and annually made the lengthy trek to Jerusalem to celebrate the prescribed feasts (Lk. 2:41). In the only narrative we have of Christ's childhood, He is lost by His parents only to be found in the temple where He asks them, "Did you not know that I must be in my Father's house?" (Lk.

---

4. John Murray, *Principles of Conduct: Aspects of Biblical Ethics* (Grand Rapids: Eerdmans, 1957), 33.

5. Anthony A. Hoekema, *Created in God's Image* (Grand Rapids: Eerdmans, 1986), 66, 68.

2:49). During His earthly ministry, it was "his custom" to go into "the synagogue on the Sabbath day" to publicly teach (Lk. 4:16). Sabbath services in local synagogues consisted of singing, prayer, the reading and preaching of Scripture, and a benediction. Jesus's consistent practice was to gather with other Jews on God's day to engage in these elements of worship. He also traveled to the Jerusalem temple for annual feasts where He taught (Jn. 7:14; 8:2, 20; 10:23; 18:20). While Christ was strongly opposed to the corruption of God's worship in the second temple, He was not opposed to the temple itself. In fact, His opposition was fueled by His loving zeal for His Father's house (Jn. 2:13–17).

Given the redemptive-historical shift that has taken place through the death and resurrection of Christ, God's temple is no longer a physical building, but a personal building made up of redeemed image-bearers who gather in synagogue-like local assemblies. If we would be imitators of Christ, our custom must be to assemble on the Sabbath, which, given the earth-shattering significance of Christ's resurrection, has shifted to the first day of the week. The Christian that is physically able to gather and chooses not to in favor of a disembodied and digitized religious experience is failing to imitate the incarnate Christ's devotion to the embodied and localized assembly.

But what about those who are unable to physically gather? Though less-than-ideal, might livestream be the best avenue for them to imitate Christ and keep the Sabbath? You could draw that conclusion. Though the screen provides an artificial experience of "a holy convocation," it provides an experience nonetheless and could serve to assist the homebound in keeping the Sabbath holy. Given the arguments throughout *Sacred Streaming?*, however, I would suggest there is a far better avenue for shut-ins to imitate Christ than by livestreaming a service. What is it? Calling upon the elders and the saints to come to them for worship and fellowship. This is what the church has historically done, and it eliminates the illusion that the assembly can be experienced and enjoyed without assembling. If you can't go to church for a prolonged season, beckon the church to come to you, and then seek to fill the rest of the Lord's Day with the word, prayer, and praise, perhaps even listening to or reading a recorded sermon or two.

The Law Model

The second ethical model looks not to the character and activity of God in Christ but to the commandments of God summarized in the Ten Commandments. This model has necessary overlap with the imitation model because the law reveals God's righteous character and was perfectly fulfilled in Christ's life. Having considered the Fourth Commandment under the imitation model, we will focus our attention here on the Second Commandment. As we saw in chapter one, the Second Commandment reveals God's fiery zeal for *how* He is worshiped. The Lord whom we worship is the Lord of His worship.

Eric Liddell, who sacrificed his dream of running in the Olympics in order to keep the Sabbath holy, memorably remarked, "The kingdom of God is not a democracy. It's a benevolent dictatorship."[6] I'm thankful to live in a representative democracy, but when it comes to the local church, Christ is King. He calls the shots in His worship, determining the place wherein we engage with Him (i.e., the local assembly), the means by which we engage with Him (i.e., the elements of

---

6.   Eric Liddell, *Chariots of Fire* (20th Century-Fox, 1981).

worship), and the manner in which we engage with Him (i.e., affectionate, whole-person devotion).

We need to take care that we do not embrace forms that have great potential to undermine the place, means, and manner of public worship He has prescribed, lest we violate the King's will and undermine His authority, setting ourselves up as functional gods in His place. Edward Fisher comments that the Second Commandment forbids "whatsoever worships are instituted by men or do *any way* hinder God's true worship."[7] Ample evidence has been provided in these pages to demonstrate that the form of livestream does at least have great potential to hinder God's true worship in some significant ways. While I have steered away from stating matter-of-factly that this form violates the regulative principle, it certainly raises serious alarms as to potentially violating such.

Of particular weight at this point is the place God has willed for His worship. It's the main reason I am comfortable streaming our services in our nursery but am not comfortable providing streaming for the homebound. For the one presupposes assembling (i.e., you can't utilize it without gathering with the local church), while the other precludes assembling (i.e., the only reason you utilize it is because you are not gathered with the local church). I personally cannot get past the unqualified essentiality of the assembly in light of the Second Commandment.

You are free to disagree, but if after significant prayer and reflection you as an individual or a leadership body are uncertain about the fittingness of livestream in public worship in light of the Second Commandment, it is best to refrain from

---

7. Edward Fisher, *The Marrow of Modern Divinity* (Scotland: Christian Focus, 2009), 284, emphasis mine.

using it lest you inadvertently stumble into idolatry and lead others to do the same. For in His law, God makes clear that the way we engage with Him in His worship is not a matter to be approached flippantly.

The Wisdom Model

In general, our relationship to technology falls within the realm of wisdom. Christ did not give us an example of how to use an iPhone, nor does the law explicitly address things like screen time. Whether or not to use certain digital innovations, along with when, where, and how to use them, require deep reflection as we apply general biblical principles to matters of great complexity. In general, technology usage defies black-and-white dogmatism. Similar to deciding at what age to stop using the rod of discipline requires parents to consider a number of factors and to draw a conclusion that might differ from other parents, so too every individual, family, and church has to think deeply and draw their own conclusions about the digital.

According to Gordon, the fundamental question wisdom asks is, "What is the likely outcome of this decision?"[8] When it comes to livestreaming church, we need to consider not only the effect it will have on us and on our present congregation, but also how it might positively or negatively influence our lost neighbors and the coming generations.

Few considerations are more weighty for me when thinking about livestream than my three sons. I want to make ecclesiastical decisions in the present that will set them up to ardently cherish the church and her worship and to pass on that devotion to future generations after I am gone. When I think about the digital mess of my sons' generation, it is a no

---

8. Gordon, *Choose Better*, 53.

brainer to me that the best thing I can do as a father is to reject any semblance of the idea that the embodied assembly can be had virtually. For their sake and the sake of my grandchildren, I feel impelled to take a clear stand now, even if that stand is deemed extreme by some.

According to the law model you might draw the conclusion that ecclesiastical streaming is permissible, but even if you do, that doesn't make it wise. What is lawful may not be helpful (1 Cor. 6:12).

Certain members of my congregation will watch the services of other churches via livestream when they are providentially hindered from gathering on the Lord's Day. My personal conviction is that this is not helpful, and I encourage my people not to do this (given the manifold reasons in chapters 2–4). But I would never rebuke a homebound church member for livestreaming another church's service or declare a dogmatic prohibition against such from the pulpit.

If you are homebound, I grieve with you over your inability to assemble with the saints, and I understand why you might choose to tune into a livestreamed worship service. You won't find a prophetic woe upon my lips. But you will find a strong encouragement to not use this technological form without first giving careful consideration to how it might be shaping you, those around you, and those who will come after you. Livestream could provide a superficial semblance of help when, in actuality, it is significantly unhelpful. The consequences may far outweigh the benefits.

The Communion Model

God fashioned us after His likeness with the unique capacity to know and enjoy Him. His great goal in creating and redeeming us is to draw us into the personal, interpenetrating, and intensely affectionate communion that the three Persons of the Godhead enjoy essentially and eternally. Every ethical decision, including the use of ecclesiastical livestream, serves to either foster divine-human fellowship or to frustrate it.

John Owen describes communion with God as "that mutual communication in giving and receiving, after a most holy and spiritual manner, which is between God and the saints while they walk together in a covenant of peace, ratified in the blood of Jesus."[9] Fellowship entails a covenantal dialogue (two-way communication), and nowhere is this more clearly displayed and experienced than in public worship as God speaks to us by word and sacrament and we respond to His gracious communication by speaking back to Him in prayer and praise. The elements of public worship are all about communion between God and His people. But does the medium of livestream help or hinder such?

The answer is obvious when it comes to the homebody who, although able to gather, willfully exchanges the public assembly for a digital one. By neglecting the bodily assembly God prizes and choosing the lightweight medium of a screen to engage in worship, he is bound to hinder communion. For livestream makes it far more difficult to receive God's communication via word, given the way this medium inherently trivializes whatever message is being spoken and opens the door for manifold distractions. It also makes it impossible to receive God's communication via sacrament. Furthermore, given the corporate and embodied nature of our prayer and praise, livestream is a sure way to hinder the church's communication back to God.

9. John Owen, *Communion with God*, in *The Works of John Owen*, ed. William H. Goold (Edinburgh: Banner of Truth, 1965), 2:9.

Those who have tried singing alone during a livestreamed service understand this. If our worship is "reactive communication" with God,[10] then we are bound to hamper such when we choose virtual worship above in-person worship.

But the answer is less obvious when it comes to the homebound who are unable to gather. Hearing the word read and preached, even if though a laptop, could help our communion with God more than not hearing it at all. So too with being virtually led in prayer and praise. But given how the Scriptures encourage us to lament when providentially separated from God's house of worship, I would suggest that using livestream when homebound may very well be frustrating your communion with God by dulling your spiritual senses to the agonizing distance through an artificial closeness. It could prevent you from grappling with God's communication in inspired laments like Psalms 42 and 43 and from fully communicating such lament back to God in prayer and song. You may be trading the limited benefits of fellowship with God through a livestreamed service for the inestimable benefits of fellowship with God in the school of lamentation which is one of God's chief means to shape His saints for their perfect communion with Him in the world to come. This is one reason why I encourage the homebound to find ways to cultivate fellowship with God that don't subtly communicate that one can partake of public worship in private like a livestreamed service inherently does.

The Warfare Model

Satan has been sabotaging God's worship and undermining God's temple from the beginning (Gen. 3:1–6). Every ethi-

---

10. Daniel I. Block, *For the Glory of God: Recovering a Biblical Theology of Worship* (Grand Rapids: Baker Academic, 2014), 25.

cal choice we make must consider his devices and how to best counter them.

There is nothing the enemy despises more than the true worship of God, and there is no true worship he despises more than public worship as the saints collectively gather to give God glory. The church's worship is not a reprieve from spiritual battle, it is the very activity that elicits the most fiery darts from the kingdom of darkness. If Satan is not successful in tempting us to stop engaging in public worship altogether, he will tempt us to a dead formality or a digital artificiality. Satan has used the digital artificiality of livestream to get a foothold into the hearts of many professing believers. I've watched as the use of this technological medium has subtly eroded convictions regarding the church and its worship in some professing Christians until those convictions become non-existent.

Gordon is on to something when he states that "cultural habit" may be the greatest tool the devil uses to deceive us.[11] When a particular behavior is normalized in a culture, it is all too easy for the people of God to be slowly shaped to embrace it as normal even if it goes against the norms established by God. The trading of in-person events for virtual events has become a cultural habit and is likely to only become more so as digital technology advances, but we must strategically war against the belief that this is normal, ideal, or appropriate, especially in the church.

Elders need to be particularly alert to the spiritual warfare that wages around the technological mode of livestream. Might this be a trojan horse by which the gates of hell are seeking to prevail against our local congregations? If so, how can we lead

---

11. Gordon, *Choose Better,* 111.

our people to strategically war against this particular device of Satan, both offensively and defensively?

More than this, we need to beware of the ways our enemy would seek to divide us over our different perspectives on ecclesiastical streaming. If after reading this book, you become convinced that your church should not publicly stream their services and your elders disagree, that is not grounds to find a new church. You can follow your convictions about livestreaming even if your church does not. If you are a church leader in a leadership body with diverse opinions about how to approach livestream, labor to come to a mediating position. Maybe one elder thinks public livestream should be available for all, while another thinks it should be available for none. Could you perhaps agree to only make your streamed services available to shut-ins and the sick through a password-protected link? I understand neither party will be entirely happy with that, but we need to realize that if we allow these matters to divide us, the devil will be overjoyed!

Furthermore, if the elders of a local church become convinced of the need to reverse course regarding livestream, they need to take great care to do so with the utmost wisdom, care, and humility. It is true that sound theology needs to be in the driver's seat when it comes to the decisions we make surrounding public worship. But sitting beside our theology in the passenger's seat must always be a pastoral-heart. Ecclesiastical decisions affect real people. So the elders need to take care to explain clearly why they believe this course reversion is necessary, to honestly welcome questions and concerns, and to tenderly shepherd the homebound whom it will most directly impact. Otherwise, a decision intended to honor the Lord could become the devil's means to snuff out the smoldering wick of a suffering saint or to foster divisive confusion in the congregation.

How Will You Respond?

We began with an industrial trash compactor closing its metal jaws upon a pulpit, a communion table, a stack of Bibles and hymnals, and a body of loving saints until those precious ecclesiastical realities were crushed into a sleek iPad Pro. If that fictional advertisement didn't lead to an audible groan when you read the first pages of *Sacred Streaming?*, my prayer is that it leads to one now as you reach the final page.

As a watchman on the walls, I have sounded my proverbial bullhorn with all the gusto I possess against this technological danger. Now you are responsible for how you respond to the fourfold warning against (1) an illegitimate authority, (2) an inessential assembly, (3) an immaterial anthropology, and (4) an incurable apathy. We might not see eye to eye on every detail of these four dangers, and we won't all respond to them in precisely the same way. But every Bible-believing, Christ-adoring person ought to be able to agree that these are four very real threats facing the church through the novel and individualistic technological medium we call livestream.

I can't tell you exactly how you must respond, but I can tell you that you must respond. You cannot afford to ignore these warnings. For the sake of your soul, the broader church, the lost world, and the coming generations, you must act in the way you believe to be most worthy of the infinitely worthy God, declaring with the apostle, "Now to him…be glory in the church and in Christ Jesus throughout all generations, forever and ever. Amen" (Eph. 3:20–21).